In 1987 scientists and engineers were seized with excitement at the discovery of 'high temperature superconductors'; these new materials become superconducting at temperatures four times higher than any previously known superconductor. Suddenly all kinds of applications of superconductivity, from magnetically levitated trains to lossless power lines, became possible.

As a result of the intense media coverage of these discoveries, superconductivity has become almost a household word, although most people have only a vague idea of what it is. In this book Professor Vidali describes in plain, non-technical terms how conventional superconductivity was discovered 80 years ago, why it took nearly 50 years to understand it, and the physical explanation of why it exists. He chronicles the developments that led up to the discovery of high temperature superconducting materials, and describes the excitement generated by announcements of the new discoveries in 1987 at a scientific conference that became known as the 'Woodstock of physics'. Finally, he speculates on possible future applications of these new materials.

This book will fascinate layman and scientist alike. Anyone interested in a clear, non-technical account of high temperature superconductivity will find it of great interest.

SUPERCONDUCTIVITY: THE NEXT REVOLUTION?

SUPERCONDUCTIVITY: THE NEXT REVOLUTION?

Gianfranco Vidali

Associate Professor of Physics, Syracuse University

CAMBRIDGE
UNIVERSITY PRESS

PHYSICS

Published by the Press Syndicate of the University of Cambridge
The Pitt Building, Trumpington Street, Cambridge CB2 1RP
40 West 20th Street, New York, NY 10011-4211, USA
10 Stamford Road, Oakleigh, Victoria 3166, Australia

First published 1993

Printed in Great Britain at the University Press, Cambridge

A catalogue record for this book is available from the British Library

Library of Congress cataloguing in publication data
Vidali, Gianfranco.
Superconductivity: the next revolution?/by Gianfranco Vidali.
p.   cm.
ISBN 0-521-37378-6 (hc). — ISBN 0-521-37757-9 (pb)
1. Superconductivity.   I. Title.
QC611.92.V53   1993
537.6′23—dc20   92-23185   CIP

ISBN 0 521 37378 6 hardback
ISBN 0 521 37757 9 paperback

UP

CONTENTS

PREFACE

Why this book?

Is superconductivity going to be the next technological revolution?
Will it have a tangible impact on our lives comparable to the
inventions of the transistor and integrated circuits? Are we
witnesses to a privileged moment in the history of science and
technology? Or will this excitement, to which the media, oddly
enough, have contributed little in comparison with the scientists'
own enthusiasm, die away as it has in many other cases? Above all,
what is superconductivity?

Naively, it would be logical to think that scientists, especially
those working on superconductivity might best answer these
questions; however, it is known that scientists enjoy speculating. It
is impossible to make an exhaustive list of scientists' opinions
regarding the impact of superconducting technology on our day-
to-day lives. Today's opinions are often in stark contrast to those
voiced just three or four years ago! In fact, we still don't know how
high temperature superconductivity really works. How can we take
advantage of a technology whose underpinnings we don't yet fully
understand?

One of the points often overlooked in such discussions is that
what appears to be scientifically possible is not always commerci-

ally realizable. An invention, even if superior in some technical way to pre-existing ones, might not be profitably produced or marketed for reasons which have little to do with science or technology. For example, high production costs or adverse environmental impact might stifle products or even technologies (the fate of the solar energy and nuclear industries in the United States comes to mind, although for different reasons).

Empowering the readers

Instead of reciting others' speculations, we throw the ball into the readers' court. By providing enough information and explanations we hope to enable readers to form their own opinions. Actually, the reader will get something extra from reading this book. As new events in superconductivity unfold (and we 'guarantee' they will), readers will be able to put them in perspective and judge for themselves the likely impact of a news-breaking discovery. While this book might become outdated, the readers will never be.

There are no convoluted, jargon-filled, obscure explanations. In fact, we require only some background in high school or introductory college physics and, most of all, a *keen curiosity*. We hope that by the time the readers reach the end of the book they will have received enough clues to speculate intelligently about these issues.

More specifically, we aim to give our readers an appreciation of the physical phenomena related to superconductivity and to illustrate how this knowledge (which is far from being complete and, at times, even satisfactory) has already affected and will continue to influence technological progress. We will not enumerate all the 'gee-whiz' gadgets that have been or will be shortly made, nor string together amusing (and often inaccurate) anecdotes, nor repeat mass-media accounts of recent discoveries. Instead, we shall focus on how scientific discoveries flow or sometimes leap from one to another. We shall start with the first discovery of the resistanceless flow of an electrical current 80 years ago, and proceed to examine events up to the latest developments in new high temperature superconducting materials.

Obviously, the task is easier when considering events of long ago, since we can distinguish between the discoveries or ideas which were seminal to a comprehensive understanding of the phenomena and those which were irrelevant. For more recent events, ones stemming from the discovery of high temperature superconductors in 1986 and 1987, we are not yet in a position to know which ideas will be fruitful and which will not, although in the past four years some trends (discussed later in the book) have certainly emerged.

Scientists and superconductivity

We are, indeed, fortunate to have witnessed the discovery of high temperature superconductors, to have been front-seat spectators, as it were, in this race towards higher and higher critical temperatures. The excitement in the laboratories is real.

While public awareness of scientists' enthusiasm has been awakened by the reporting of recent events, such excitement is not new in science; it is often found whenever a significant discovery is made or an understanding of a complex phenomenon is reached. It is an excitement comparable to the one felt by someone who has worked for a long time at a complex puzzle or riddle. All of a sudden all the gathered pieces which seemed somewhat important before but couldn't be placed in any sensible order, fall into their proper places. The riddle is finally solved and the sensation of accomplishment is overwhelming ('I've got it!!, Eureka!!'). The joy comes from the realization that we can see not only the meaning of each single piece, but can recognize a design which, until a few moments before, had been hidden from us.

In the realm of the natural sciences, most of the joy and euphoria comes from the discovery of, as many scientists say, 'how clever Nature can be.' One of the most important goals of this book is to try to catch this excitement and involve the reader as a participant in this joy. We hope to accomplish this by providing readers with an understanding of how the discoveries came about and what they meant, rather than through involvement as passive spectators of a newsreel of soon forgotten facts.

Acknowledgments

This book couldn't have been completed without the help of Carole, my wife, who spent countless nights reading, criticizing and editing the manuscript. I would like also to thank Dr Simon Capelin, senior editor of Cambridge University Press, for his constant encouragement.

This work was completed while I was spending my sabbatical leave at Princeton University. The hospitality of the Chemistry Department is gratefully acknowledged.

G. Vidali

Syracuse, 1992

1

Introduction and overview

1.1 March 1987

'THE FASCINATION OF SUPERCONDUCTIVITY is associated
with the words perfect, infinite, and zero', muses Brian Maple,
Professor of Physics at the University of California, San Diego.

In the language of superconductivity *perfect* is the expulsion of
magnetic fields from a chunk of material that has become
superconducting, *infinite* is the electrical conductance, and *zero* is
the electrical resistance. Perfect and infinite are words that any of
us might use many times a day often absent-mindedly. We might
talk excitedly of 'perfect parties', although we realize that, on
second thought, we have attended even better ones. And we boast
of having 'infinite patience' before we recognize it is often tried
beyond its limits. Everybody understands this, and seldom is
anyone asked to explain what perfect and infinite really mean. But
scientists, and physicists in particular, when not at parties or
shouting at their children, have different ideas about 'perfect,
infinite and zero'. In fact, if they belong to that group that likes to
tinker in basements of university buildings – more correctly called
'experimentalists in laboratories' – they are especially reluctant to
use such words as perfect, absolute, and infinite. When talking
about facts of science, they have been trained to disregard such

I

words, since nothing can be made perfectly, nor be measured infinitely large or small. When discussing superconductivity, however, physicists go wild and use these and other rather hyperbolic adjectives quite freely and without hesitation. Why?

Let us turn the clock back a couple of years. New York City, 1987.

'Where do you want to eat?' asked a couple of people in the group, almost simultaneously.

'We better find something quick. It's gonna be pretty crowded.'

At about 5:30 p.m., midweek, on a crisp March day in mid-Manhattan it shouldn't have been a problem to find a place to eat. But we, a loosely bound group of five or six physicists attending the American Physical Society Meeting, knew we had to find a reasonable place to have dinner near the Hilton, where our meeting rooms were located. We knew that three thousand or so colleagues were on the same hunt at the same time.

Time was indeed the issue. At seven o'clock, a special session was going to be held in which new results about the new superconductors were to be presented. Many speakers were scheduled to give short presentations of their latest work in superconductivity, and results which had not yet appeared in print were likely to be announced.

When we arrived at the hotel, about half an hour before the scheduled opening of the session, we immediately realized that we didn't need to ask for directions. A huge crowd of people, discussing animatedly, was pressing at the doors, still closed, of the hallway leading to the big ballroom where the meeting was about to begin. When the doors opened, a rush to the empty chairs quickly ended and most people had to find much less comfortable positions on the floor around the perimeter of the large room or in the hallway (Figure 1.1). A long night had begun.

For hours, group leaders from famous scientific institutions as well as lone scientists from small universities and colleges went to the microphone and transparency projector to illustrate their new discoveries. TV camera crews and photographers were constantly moving around often creating a human wall in front of the speaker. But with some ingenuity and luck, one could grasp most of what was said from the podium. New superconducting materials! New

Figure 1.1   The Woodstock of physics: the special session on superconductivity held during the 1987 March Meeting of the American Physical Society (Hilton Hotel, New York City). (Courtesy of the American Institute of Physics Niels Bohr Library.)

mechanisms for superconductivity! Unheard-of transition temperatures! Here is a sample of a superconducting tape! Everybody was excited. It was an important moment in the history of twentieth century science.

That was March 1987. What of those events is left today? What have we learned in four or five years? Should we still be excited about these high temperature superconductors? Do we understand how they work? What do scientists and engineers say about them? And whatever happened to the superconductivity gadgets people were talking about a few years ago? Superconducting trains, cars and power lines: are they being built? And what is superconductivity, anyway?

1.2 Superconductivity before 1986

Perhaps the time has come to consider and reflect upon the events of 1987, when superconductivity not only became a hot topic of conversation at scientific institutes around the world, but was introduced to a great many people who had never even heard the word before.

In fact, many don't know that superconductivity was discovered 'long ago', more than 80 years ago by Kamerlingh Onnes at the University of Leiden, in the Netherlands. Since then researchers at universities and in industrial laboratories have spent countless hours trying to find those special materials that, when cooled below a certain temperature (called the *critical* or *transition* temperature) possess those spectacular characteristics which we now associate with the name superconductivity. There are two properties of

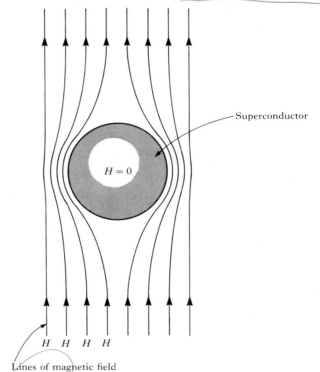

Lines of magnetic field

Figure 1.2   A superconductor expels an external magnetic field. Lines and arrows show the direction of the magnetic field.

superconductors which cannot be easily missed. First, when a chunk of matter becomes superconducting, the resistance to the passage of an electrical current goes to zero. It doesn't just become very small, it really becomes zero. This is a very remarkable property. Electric power is expensive to produce, as we all know; it is also expensive to transport since to overcome this resistance some additional power has to be provided. Every trick has to be used to minimize losses since electric power has to be transported for hundreds if not thousands of miles. Even using the best conductive materials available, such as copper, an appreciable amount of power is lost in overcoming this resistance. Other tricks to minimize losses, such as carrying electricity at very high voltages (100 000 volts or more), often present drawbacks, such as an adverse environmental impact.

The second property is less apparent than the first, but not less important. Let us take the same chunk of material and place it near a magnet. The magnet will exert an influence on this piece of matter; we will call this influence 'the magnetic field' and this has been known since antiquity. By the way, we could prove that there is indeed something there by moving a compass needle near the magnet; the compass will align its needle in certain directions (called magnetic field lines) pointing toward (or away from) the pole of the magnet. In short, we will say that our chunk of material is in a magnetic field. In most cases, experimentation shows that the magnetic field will penetrate the sample and exit from the other side. At ordinary temperatures, unless this chunk is ferromagnetic, such as a piece of iron, nothing remarkable will happen. However a remarkable thing occurs when this chunk is cooled down; all of a sudden the material expels the magnetic field from its interior. The magnetic field lines which went through the sample are now rerouted to its periphery (Figure 1.2). In other words, the inside of this sample becomes shielded from any magnetic force. It turns out that the temperature at which the magnetic field is expelled is the same as that at which the sample shows no electrical resistance. If the sample is cooled down further, these two properties remain (and we say that the matter is in a 'superconducting state'); if the sample is warmed up above this temperature, these effects disappear (and the matter is now in the 'normal state').

Contrary to popular belief, superconductivity (to be defined for now by the two effects mentioned above) is not as rare as it might have seemed at first, since many metals (and other types of substances as well) are indeed superconductors. The fact is that most superconductors become so at what most people would consider extremely low temperatures, much colder than any temperatures measured on Earth (except in laboratories!). At the beginning of this century, Onnes was the only one who had the technology to reach these low temperatures. In fact, he had found a way to produce the coolant needed for reaching these temperatures, liquid helium. Thus, it was just a matter of time before one of the most striking properties of superconductivity (specifically: infinite electrical conductivity) was discovered in a number of materials.

Kamerlingh Onnes soon realized the practical significance of this amazing discovery. Electrical power could be transported without appreciable loss over great distances, and electromagnets, in which the magnetic field is produced by an electrical current, could be built that would have magnetic fields otherwise unattainable. To Onnes's great disappointment, this discovery couldn't be applied as easily as he and his contemporaries first thought. He soon found that a modest magnetic field, comparable to the one generated by a small household horse-shoe magnet, could easily make the material switch from a perfect conductor to an ordinary material with a finite, instead of infinite, electrical conductivity. Onnes also found that an electrical wire of a given size, when in the superconducting state, could carry electrical current only up to a certain limit before resistance to the passage of the electric current set in. As before, the 'superconducting state' is destroyed and the material acquires back its 'normal' electrical properties. To be sure, this resistance can be exploited for useful purposes. Electrical resistance causes an electrical current to dissipate heat and this phenomenon (Joule heating) is used to run hot water tanks or baseboard heaters. However, in many other instances, such as the transport of electrical power, the resistance should be minimized, or some electrical power will be spent in unwanted heat dissipation. Furthermore, the maximum temperature at which these materials still showed superconducting properties didn't go much higher than the astoundingly small temperature 4 K (K is for Kelvin or about -452 °F) of the original discovery of superconductivity in mercury wires. Four degrees Kelvin is a really low temperature; it is about the temperature found in deep space. Not surprisingly, it is difficult and expensive to cool materials to these temperatures. For superconductors to be used on a large scale for their exceptional properties, their operating temperatures need to be much higher.

In seventy years or so of intense research, many more materials were discovered to have superconducting properties. However, notwithstanding the effort of many scientists and engineers, the maximum temperature at which materials still showed super-conductivity (the critical temperature) moved up very slowly over the years, to reach 23 K in the early seventies.

If progress in raising the critical temperature was slow, so was progress in reaching an explanation for the odd behavior of superconductors. For a long time the best minds in science were challenged to find out how superconductivity really worked, but with little success. Finally, about forty years after the first discovery in Onnes's laboratory, a mechanism was found which could account for all the superconducting properties. Soon theoretical accomplishments fed engineering efforts to design new devices exploiting superconductivity. Progress was rapid, and many new instruments and machines came off the drawing boards. Powerful superconducting magnets produced the magnetic fields necessary for the now nearly ubiquitous Nuclear Magnetic Imaging instruments, in which an image of the body tissues is obtained for medical diagnosis. Superconducting wires were used in the construction of powerful magnets to bend subatomic particles in circular accelerators (as at Fermilab and the still-to-be-built Superconducting SuperCollider) to study the ultimate constituents of matter. Superconducting magnets were used in magnetic levitated trains being tested in Japan. Other superconducting properties, to be discussed later, were exploited to make the very fast electronic devices which will be needed in future super-computers and scientific instruments.

However, the same theory which explained superconductivity predicted that it was unlikely to find a material which could have a transition temperature much greater than 30 K. Superconductivity at room temperature was confined to science fiction books.

This was in essence the status of superconductivity in the mid-

80s. Then, in 1986, the world of superconductivity research and engineering changed drastically and the repercussions will be with us for a long time to come. Let us look at how this happened.

1.3   Superconductivity after 1986

Georg Bednorz and Alex Müller quietly reported their results of a 'possible' critical temperature of 30 K in a European physics journal in 1986. A critical temperature of about 30 K was undoubtedly higher than the 23 K which was the highest one previously known, but was still far below the temperature at which commercialization of superconducting devices would be feasible (well above 80 K, as we will see later). Thus, it is understandable that this result was under-reported and went largely unnoticed by the general public. However, it didn't go unnoticed among Bednorz and Müller's colleagues and other scientists. Among the oddities of their findings was the type of material used, a type of copper oxide (a material with oxygen and copper atoms bonded together), which hadn't up to that time been considered one of the best possibilities for a high temperature superconducting material.

When Paul Chu, Maw-Kuen Wu and collaborators found superconducting materials at much higher temperatures (about 90 K) at the beginning of 1987 the world took notice. At the celebrated 1987 March Meeting of the American Physical Society in the New York City Hilton a great excitement overtook the thousands of scientists who were there (Figures 1.3 and 1.4). A colleague from my institution was at that time talking to a fellow physicist in the press room. He observed how extraordinary this meeting was, calling it a *Woodstock of physics*. Perhaps a reporter heard him. At any rate the '*Woodstock of physics*' nickname appeared in newspaper headlines across the country. For a few months in the spring of 1987, hundreds of scientists, Nobel laureates as well as the greenest of graduate students, went on a work rampage. It was indeed a race, a race to obtain fame and patents. In our exposition we will see that this rivalry was not much different from the one that plagued scientists, or natural philosophers as they were then called, almost a century earlier. With respect to eighty or even just twenty years before, striking differences existed: the number of people involved, the equipment

Figure 1.3   Some of the protagonists of the special session on superconductivity (March 1987, APS Meeting, Hilton Hotel, New York City). From left to right: Alex Müller, Paul Chu, and Shoji Tanaka. (Courtesy of the American Institute of Physics Niels Bohr Library.)

in the laboratories, the FAXes, Federal Express packages, the lawyers, and the videocameras made the 1987 event so fast, so loud and so frenzied. But after all the dust had settled, the results stood up against worldwide scrutiny. In the four years since that March Meeting, considerable progress has been made (though with less fanfare) and high temperature materials have found their way into commercial products, although they haven't found application yet in the magnetic levitated trains and electric powered cars once heralded as being 'just around the corner'. However, progress in this field is very rapid. Notwithstanding the difficulties inherent in fabricating superconducting materials with the characteristics desired for industrial processing, and despite the initial disappointment with some of the superconducting characteristics of the new materials (for example, low electrical current density capacity, as we will see later), there are already commercial products on the market based on high temperature superconducting properties. It is reasonable to assume that it will be just a matter of time before engineers find a way around some of the 'undesirable' properties of high temperature superconductors. Or, perhaps, they will discover some other new useful characteristics!

1.4   How things work

The discovery of superconductivity was the more remarkable because it was totally unexpected even by those very scientists working in the field. What led them to it? How will high temperature superconductivity be exploited? After the age of

Figure 1.4   Another protagonist of the 1987 APS March Meeting: Georg Johann Bednorz, winner, together with Alex Müller, of the 1987 Nobel Prize in Physics for their discovery of high temperature superconductivity. (IBM, courtesy of: AIP Meggers Gallery of Nobel Laureates.)

semiconductors and computers, are we on the verge of a new age, the age of superconductivity?

Understanding how things really work is just as important as inventing a new technology. Sadly, the idea of figuring things out is more associated with painful memories than with the joyful ones it ought to be. It is fair to say that while some people's memories of high school or college physics are vivid simply because their experiences with the subject were so horrendous, many of these same people like the challenge of 'figuring things out', whether it is a crossword puzzle or putting back together a toaster (without leaving extra bolts and nuts on the kitchen table). Physics is the book that tells us how all the pieces of the puzzle that Nature presents to us on an enormous table seem to fit together. Oh yes, sometimes She doesn't give us all the pieces and sometimes She throws in a few extra ones, but, overall, we have come a long way toward understanding how many things work and, most importantly, how new things *will* work. Perhaps the reason why physics has its undeservedly bad reputation (we mention physics because people seem to dislike it much more than other sciences, such as chemistry or biology) is that, in many cases, physics is presented as a collection of facts which, at a first reading, do not seem to be connected much at all to the real world. Most of what we learned in school about physics is connected with improbable massless and frictionless pulleys and springs. Indeed it seems that introductory physics textbooks will never teach us how 'real' things, such as bridges, transistors, etc., work.

Superconductivity is strikingly different from springs, pulleys and tops, not only because its effects are so evident even to the public at large (electric currents flowing forever!) but because the explanation of it is so bizarre that it was missed by the best minds for decades. It is not much of an exaggeration to say that figuring out how superconductivity worked was perhaps one of the most difficult and finest achievements in the history of modern science.

1.5   The tour

Superconductivity is just the right subject to present in the order of the events as they unfolded, from the first experiments of Onnes, to the long efforts to explain it, and finally to its commercial exploitation just before new high temperature superconductivity materials were discovered. It will become clear that Nature didn't hand down the solution right away; people reached it by a combination of brilliant intuition, tedious but necessary hard work, serendipity and, yes, good luck. Once we grasp how the 'old', low temperature superconductivity works and has been exploited, we will then look at the new high temperature super-conductors. We will recount how Bednorz and Müller were led to this discovery and the different avenues of research pursued since in many other laboratories throughout the World. We still don't have an explanation of how the new superconductors work; we are not sure whether the mechanism of high temperature super-conductivity is completely different from the old one at low temperature. But there is a very intense effort by theorists to come up with new (and sometimes very exotic) explanations. We will look at some of these and then we will describe some of the applications of the new materials in devices, instruments and machinery.

Finally we should warn sophisticated readers that this book is not for those already versed in superconductivity. Omissions, short-cuts, and simplifications have had to be made in order to make this book readable for a wide range of people with different backgrounds and interests.

2

Physics at Leiden and the liquefaction of helium

2.1 Science 100 years ago

SUPERCONDUCTIVITY WAS BORN IN 1911 in the laboratory of Kamerlingh Onnes at the University of Leiden. While he was testing theories of how electricity is carried in metallic wires at low temperatures, he unexpectedly found that electric currents can flow in certain materials without encountering *any* resistance and without dissipating *any* heat. He and his contemporaries immediately realized the enormous potential applications of such a discovery. Even before this discovery, engineers were considering cooling copper wires – which don't become superconducting – to lower the electrical resistance of these cables for long distance transmission of electric power. Now something undreamed of happened: electrical power could be sent without any heat dissipation! The only problem was to find a suitable material that would remain a superconductor at a temperature higher than the one used for the first experiments, about −269 °C. This is just 4 degrees higher than the lowest temperature possible, −273 °C. Ordinary temperatures are around 20 °C† (70 Fahrenheit). The

† °C stands for degrees centigrade; the temperature interval between ice (0 °C) and boiling water (100 °C) is divided in 100 equal intervals called degrees. Ambient temperature is about 20 °C. To convert to Fahrenheit degrees multiply by 9, divide by 5 and add 32. In the USA the Fahrenheit scale is more widely used than the Celsius

materials used by Onnes lost their superconducting properties at temperatures of $-250\,°C$ or colder (in comparison, the coldest spots on Earth never go much beyond -50 or $-60\,°C$), making any large scale applications quite impractical. The problem of creating superconductivity at room temperatures is still not solved today, although many scientists are much more optimistic now than they were just a couple of years ago. Paradoxically, we can say that since we know so little about the new high temperature superconducting materials and how they work, the realm of possible exciting new technological applications is more wide open than ever. Onnes felt a bit the same way in his own time after the discovery we are going to describe shortly.

Why did Onnes look at the electrical resistance of metals at low temperature? Had he predicted what he was about to discover? How did he understand superconductivity? Why did the citation for the Nobel Prize in Physics awarded to him in 1913 refer to 'his investigations on the properties of matter at low temperatures which led, *inter alia*, to the production of liquid helium' without singling out his discovery of superconductivity? Was he a man well ahead of his time, as Galileo, Newton or Leonardo da Vinci had been centuries earlier, or did he precede his contemporaries by just a few years? A great deal of the fascination with superconductivity lies in the fact that its manifestations were very clear but its ultimate explanation (some forty plus years later!) was extremely unconventional and unintuitive to most physicists. Thus, it pays to spend some time recounting how Onnes (and others) arrived at their discoveries.

Physics (and science in general) is conducted in a far different way now than it was a hundred years ago. Today extensive coverage in the media, especially television (in programs such as *Nova*) and magazines (*Science, Discover, Scientific American*) has give us images and descriptions of where and how scientific research is done. We have all seen beautiful photographs of living cells, atoms in artificially made materials, and awesome pieces of complex machinery, each illuminated by terse, yet impressive, accounts of new discoveries and inventions. And many of us might

---

or centigrade scale. In the former the ice/water temperature is 32 °F and boiling water is 212 °F. It has been said that the Fahrenheit scale was invented by a madman. We couldn't agree more!

have wondered once or twice how in the past scientists, prac-
titioners and amateurs made discoveries (considering how few
and primitive were the instruments available). Actually, when one
looks at old photographs of typical turn-of-the-century laboratories (see Figure 2.1, for example), one wonders what could
have been measured, since quite primitive equipment and few
instruments were available.

Heike Kamerlingh Onnes was very well aware of the importance
of having the right tools and people to do research. In 1904 he
wrote:

Many improvements have been made at Leiden during the last two
decades ... Remember the first gasmotor of which the fly-wheel often
would not move unless the director and the assistants had worked
themselves out of breath, and also the gloomy small cellar with a
wooden lathe for the instrumentmaker.

We will see later that some of these differences were not
superficial, but reflected the fact that scientists were educated
differently then; science played a different role in society too. Yet,
Onnes saw that things had to change in order to carry out research
that was becoming more complex and expensive every day. He
wrote:

The instruments which in the cryostats are brought to the required
temperature must in most cases be especially constructed for this
purpose. This is another reason why we cannot succeed in making
accurate observations at low temperatures unless we devote ourselves
entirely to it – unless we specialize. Indeed the time is not far off when
in all laboratories the physicists will choose a special line of
investigation.

Parenthetically, it is interesting to note that this is still true today
if we are striving to do the best research, although many
apprentices in the sciences might think that the reverse, that is,
having a 'turn-key' system or off-the-shelf instrumentation, is the
most desirable approach.

Considering the laboratories and facilities they had (no computers, hand-held calculators, Xeroxing etc.), how did these men
and women accomplish so much? What did they know? What were
the clues which led to their discoveries?

To answer these questions we have to take a trip into the past.
Our voyage starts with a survey of the status of low temperature

Figure 2.1(a)   Kamerlingh Onnes in his laboratory; van der Waals is on the right. (Courtesy of the Kamerlingh Onnes Laboratorium, Leiden.)

Figure 2.1(b) Apparatus in Onnes's Laboratory. (Courtesy of the Kamerlingh Onnes Laboratorium, Leiden; copyrights: Elsevier Science Publishers.)

physics in Europe in the late 1880s, about a quarter of a century before the discovery of superconductivity.

2.2   The quest for the liquefaction of gases

In the 1880s, the place to be to make discoveries in the physical sciences was not an Ivy League school in the United States but the laboratories of Northern Europe, such as the Academy of Sciences in Paris or the Royal Institution in London. At that time, a number of scientists were busy trying to liquefy gases, i.e. substances which at ordinary temperature and pressure are in the gaseous state, such as nitrogen (which constitutes 80 % of the air we breathe), oxygen, hydrogen, etc. After some experimentation with liquefying gases at progressively lower temperatures, scientists began to envision building a machine that, by using a cascade of liquefaction processes with different liquefied gases at different starting temperatures, could reach the lowest temperature attainable, i.e. -273 °C. Scientists found it convenient to define this lowest temperature as 0 degrees. 0 °C equals 273 in the new scale, called the absolute or Kelvin scale (after Lord Kelvin as elaborated below) or K for short.†

The fact that there *is* a minimum temperature was realized in the late seventeenth century when it was observed that a drop in the temperature of air was accompanied by an equal drop in pressure. It was easy to calculate, from data taken then, that when the pressure becomes zero the temperature reaches a limit, -273 °C, or equivalently, 0 K.

Later, it was realized that the state of 0 K couldn't be reached by any machine man could conceivably build, but this was not known in the 1880s.

In 1877, Caillitet was working in Paris on an apparatus to liquefy gases. His apparatus was a complicated assembly of pipes, capillaries (very thin tubes) and glass vessels in which a gas was compressed at high pressure in the belief that it would liquefy if enough pressure was applied to it (we will explain this point later). One day, during an experiment, a capillary sprang a leak and,

---

† The melting point of ice is 273 K or 0 °C; 1 K and 1 °C measure the same temperature interval, but their reference or starting point is different. Degrees centigrade are used in everyday life in most industrialized countries; degrees Kelvin are preferred by the scientific community.

under high pressure, acetylene, the gas that Caillitet and his assistants were trying to liquefy, started to escape. A mist briefly formed on the glass walls. Many might have overlooked this curious misting of the walls; instead, Caillitet realized that the misting was due to the condensation of cold acetylene vapor; he had, in effect, liquefied acetylene.

Caillitet's reasoning was correct: when a gas at high pressure expands into a region of low or no pressure (vacuum) the gas cools. This is analogous to the cooling felt by the hand when operating a spray-can for some length of time. Under appropriate conditions this cooling is sufficient to make the gas condense into liquid droplets. Soon he liquefied oxygen using the same method, thus achieving an important milestone (at ordinary atmospheric pressure liquid oxygen boils at about 90 K or -183 °C).

Can any gas be liquefied? What is the starting temperature and pressure necessary to achieve liquefaction?

Boyle's law, derived from experimental observations, was well known at the time; the law says that at a given temperature the volume and the pressure are inversely proportional. By decreasing the volume and keeping the temperature constant the pressure goes up according to the relation:

$$P = \text{constant}/V,$$

where P is the pressure and V is the volume occupied by the gas. The constant is directly proportional to the temperature, which means the higher the temperature, the higher the pressure. This can easily be verified by increasing the temperature of a pot of water with a lid on it.

Thomas Andrews, in the 1860s, observed that in certain cases when decreasing the volume the pressure of the gas didn't go up as the temperature was kept constant. This happens when gas is converted into liquid, see Figure 2.2. This is what Caillitet originally wanted to achieve. Obviously the trick of decreasing the volume works for temperatures less than T_c (called the critical temperature).† For $T > T_c$ (i.e. T greater than T_c) no condensation occurs, no matter how much squeezing is done; the pressure just keeps going up and up. The best experimentalists realized that it

† This critical temperature of gases is *not* related to the critical temperature of superconductivity defined earlier.

was important to know the P–T (pressure–temperature) diagram of the substance to be liquefied (Figure 2.3 and Table 2.1). Obviously different substances have different T_cs and a great deal of time can be wasted if the wrong initial conditions of temperature and pressure are chosen. It is also important to notice that the pressure at T_c is generally different, and in many cases higher, than ordinary atmospheric pressure; hence, the need to compress gases to achieve liquefaction.

Caillitet didn't actually achieve liquefaction by applying pressure as he had originally intended. Instead, he discovered another method of liquefying gases. In this method, the gas cools during expansion from a vessel at high pressure to one at low pressure. We know that pressure in such a vessel is created by random banging of atoms or molecules on the walls of a container. When gas is made

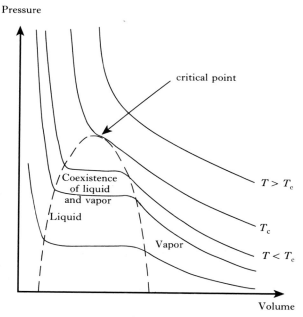

Figure 2.2   Pressure(P)–volume(V) diagram for a simple substance. Solid lines show isotherms, that is, regions of constant temperature. The region of coexistence of liquid and solid is bounded by the dashed line. No condensation of vapor into liquid occurs for temperatures greater than the critical temperature (T_c).

Table 2.1 *Critical temperatures and pressures for selected substances.*

Substance	Critical temperature (K)	Critical pressure (atmospheres)	Boiling point (K) at 1 atmosphere
Ethylene	283	50.5	169
Acetylene	308	61.6	179
Oxygen	155	50.1	90.2
Nitrogen	126	33.5	77.3
Hydrogen	33.3	12.8	20.4
Helium	5.2	2.3	4.21

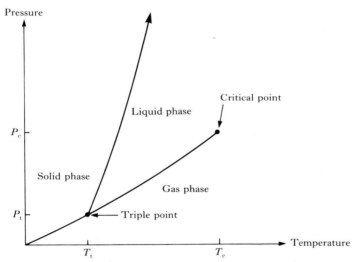

Figure 2.3 Pressure(*P*)–temperature(*T*) phase diagram for a simple substance. T_c and P_c are the critical temperature and pressure, respectively. The triple point, where the gas, liquid and solid coexist, is also indicated. Temperatures greater than the critical temperature are possible. Lines indicate phase boundaries.

to expand suddenly, most of the velocities of the particles are converted from random directions to the direction of the flow. More precisely we say that the energy of the atoms or molecules is converted from kinetic energy (the energy associated with the velocity of the particles) due to random motion to kinetic energy in the direction of the flow. Temperature is nothing but a quantity describing this randomness; if random motion decreases, the gas cools.

Wroblewski and Olszewski (University of Cracow-Poland) were also in the race to produce liquefied gases. Liquid oxygen was necessary to bootstrap the liquefaction at lower temperatures. Liquid oxygen had been produced by Caillitet but not in stable form; in other words, he was not able to collect enough liquid to be used for anything practical. The Polish scientists used a different method. Oxygen gas was put under pressure in a capillary surrounded by liquid ethylene; a vacuum pump was connected to the top of the ethylene bath until 25 torr of pressure was reached (1/30 of an atmosphere, 1 torr being 1/760 of an atmosphere). The ethylene liquid cooled down to −130 °C; the oxygen, at high pressure, started to condense.

Why does a liquid cool when its vapor is removed? The vapor on top of a liquid is composed of atoms (or molecules) which have a higher kinetic energy than the ones in the liquid: this is why they manage to escape from the liquid in the first place. By removing these molecules (with the use of a vacuum pump), we eliminate the possibility that they will condense back onto the liquid surface; if we progressively remove the molecules with the highest kinetic energy, we will slowly reduce the amount of kinetic energy of the atoms in the liquid and therefore the temperature of the remaining liquid. This fact is well understood by the nomads of the Sahara Desert who use evaporating water to keep food cool under the hottest sunlight.

The next gas to be liquefied was hydrogen, but a large jump in temperature lay between the boiling point of liquid oxygen and the suspected critical temperature of liquid hydrogen (see Table 2.1). The boiling point of hydrogen is at about 20 K or −253 °C; for the reason given above it is, however, more important to consider the critical point.

Not much progress was made until James Dewar, in London, invented a vessel to maintain a liquid at temperatures much lower than the ambient temperature for an extended period of time (Figure 2.4). The principle behind his invention is very simple; the vessel has a double wall and the air in the space between the walls

Figure 2.4 Painting showing Sir James Dewar performing a physics demonstration in a public lecture at the Royal Institution of Great Britain. (Courtesy of the Royal Institution.)

is removed. In this way, heat can reach the cooled liquid only by radiation, that is, electromagnetic waves† and not by contact or convection, that is, transport by air. This type of vessel, called a 'dewar' nowadays, is still widely used in low temperature physics. A Thermos for containing liquids, such as coffee or iced tea, is built on the same principle. With this and other improvements, Dewar was able to produce liquid hydrogen.

The competition to be the first to liquefy a gas or to reach a very low temperature was keen. There were claims and counterclaims; sometimes one group beat another by just a matter of days. Announcements of breakthroughs in attaining lower and lower temperatures were made at meetings of the various European scientific societies. Promptness in announcing new discoveries was very important. While nowadays this race might appear quaint – though it was probably beneficial since it spurred competition – in one instance it produced deleterious effects. Dewar, as the story told by the physicist Kurt Mendelssohn goes, presented a paper in May 1889 at the Royal Society about the liquefaction of hydrogen. A colleague of his, Ramsey, publicly claimed that Olszewski in Cracow had already liquefied hydrogen. The claim was not true and was based on nothing more than a rumor; Ramsey, when confronted with a disclaimer, didn't retract his position in print, and Dewar published his account of the controversy. The end result was that the relations between the two reached a low point. This was to Dewar's disadvantage, since Ramsey was one of the few people in England able to obtain sufficient quantities of the newly discovered element, helium.

Helium, a gas quite scarce on Earth, is very hard to isolate since it is inert and reacts with no other substance. Helium was discovered in 1869 by Ramsey following spectroscopic measurements of the solar corona, which indicated the presence of a new element in the sun. In fact, an element can be 'fingerprinted', or identified, by the color of light emitted when it is heated to high temperature.

† Visible light and microwaves – as used in a microwave oven – are two examples of electromagnetic radiation; electromagnetic waves interact with matter in different ways depending on the frequency of the waves. Frequency and wavelength – the length over which the wave repeats itself – are inversely related. X-rays have very short wavelengths and interact little with matter; microwaves and radio waves have much longer wavelengths and may be either absorbed by or reflected from objects.

To isolate helium on Earth was quite a difficult task. Dewar, and others, wanted to liquefy helium, since this was the last inert gas to be liquefied; it was expected that its boiling temperature would be even lower than that of hydrogen. The honor of being the first to liquefy helium was not the only motivation in this race. Dewar realized the importance of liquid helium as a tool to reach the lowest attainable temperatures. Liquid helium could be used as a refrigerant as the other liquefied gases had previously been; liquid helium would then become an essential instrument in achieving scientific discoveries about matter near the absolute zero temperature. The group that had the ability to produce it would have a clear advantage in pursuing research on properties of matter at low temperatures. Unfortunately, Dewar had very little helium and what he had wasn't very pure; often the impurities would freeze during a cool-down causing clogging of the narrow capillaries. During such a mishap, almost all the helium that Dewar had was vented out by one of his assistants. Dewar was out of the race.

2.3 Kamerlingh Onnes's laboratory

Notwithstanding his remarkable achievements, Dewar's scientific production was hampered by a lack of good technical facilities and the acrimony he sometimes showed toward his colleagues and assistants. Onnes's style was different.

Appointed to the chair of experimental physics at the University of Leiden (The Netherlands) in 1882, Heike Kamerlingh Onnes devoted a great deal of time to making Leiden the center of low temperature experimental physics. This task required much technical and managerial ability. He brought in the best glass blowers and technicians available. It was an honor to work in Onnes's laboratory. Assistants were paid, but they were required to 'donate' some money for the upkeep of the laboratory! He invited many scientists from all over Europe to spend their sabbaticals (research leaves) at Leiden. This was a far cry from the secrecy that sometimes surrounded the experiments to liquefy gases in other laboratories. All these steps turned out, in one way or another, to be beneficial for his research and the state of his laboratory for many years to come.

Onnes's philosophy emphasized the need for accurate quan-

titative measurements. His interests were not only in the liquefaction of gases, but in the properties of matter at low temperature. It
was not then by chance, the title of his keynote address in
celebration of the University of Leiden's 329th anniversary was:
'The importance of accurate measurements at very low temperatures'. He took time to design his apparatus to liquefy first air,
then helium, which he obtained with the help of his brother who
held a high ranking post in the government. He spent much time
measuring isotherms, that is, measuring the pressure of a gas as a
function of volume at a given fixed temperature (see Figure 2.2).
He studied the new theory of gases proposed by his colleague van
der Waals, a modification of $pV = $ constant $\times T$ (Boyle's law) to
take into account any observed deviations from the equation above.
In his theory, van der Waals elegantly showed how to account for
these deviations by considering simple parameters related to the
size of the gas atoms and the interaction forces between the atoms.

When his apparatus was ready and tested and when he had
collected enough information to predict its critical temperature,
Onnes and his collaborators (in particular Holst and Dorsman) set
out to liquefy helium. At that time, progress reports of on-going
research at Leiden were published in the *Communications from the
Physical Laboratory of the University of Leiden*. The nature of this
publication falls somewhere in between the technical reports from
industrial laboratories, such as IBM, Xerox etc., and the scientific
peer-refereed journals. There were many technical details on how
the apparatuses were built or experiments carried out. Often these
reports were written in a very candid way, as a writer would jot
down notes in his diary. For example, a gripping description was
given of the long day in which the liquefaction was first attempted.
The day was July 10, 1908.

It was a wonderful moment when the liquid, which looked almost
immaterial, was seen for the first time. It had not been perceived when
it flowed into the glass; its presence could be detected only when the
glass was filled. Its surface stood out sharply defined like the edge of
a knife against the glass wall. I was overjoyed when I could show
liquefied helium to my friend van der Waals, whose theory has been
guide in the liquefaction up to the end.

Onnes's laboratory was known throughout Europe. It was said
that the coldest spot on Earth was in Leiden. His attention to

detail, meticulous planning, receptiveness to the ideas of his colleagues and managerial skills made the difference. In respect of this, it is interesting to look at the transcriptions of the lecture he gave in Stockholm when awarded the Nobel Prize in Physics. He devoted most of his lecture to the liquefaction of gases; ample space was given to the description of his apparatus; numerous drawings completed the presentation, one of which is represented here (Figure 2.5). Such a style is in agreement with the image we have of Heike Kamerlingh Onnes, a man dedicated to careful experiments and observations.

Onnes had created a very impressive laboratory; his managerial and scientific accomplishments can hardly be disputed. Perhaps because he was so famous and so many people came in contact with him, such as visitors to his laboratory, assistants, technical staff, students etc., a number of stories circulated about him. It is hard to sort out the most accurate from the most embellished ones, nonetheless a picture emerges of Onnes as a formal and demanding professor. This, at least, was the opinion of another good Dutch physicist, Hendrik Casimir, who received his first training at Leiden. Although Onnes knew how to motivate his technicians, he had a paternalistic attitude toward them. Students and assistants, that is, those whom we would nowadays call post-doctoral students, didn't fare much better. Most of the publications issued by Onnes do not bear other names as co-authors; however, at the end of many papers, and in particular the ones about the early findings of superconductivity, his assistants were mentioned. One of his assistants, Holst, later became a successful director at the Philips Laboratory in Eindhoven (The Netherlands).

Much has been said about his motto: '*Door meten tot weten*', which he wanted to inscribe on the doors of each laboratory. Translated it means: by measurements to knowledge, that is, only by doing measurements can one obtain knowledge. Unfortunately, according to Casimir, who respected Onnes but didn't like his paternalistic and 'old fashioned' style, many physicists at Leiden and elsewhere distorted the original intent of the motto, that is, that careful observations and measurements are necessary to obtain knowledge about natural phenomena. In some cases it was interpreted to mean that *a lot* of measurements had to be taken, sometimes with some disregard for thoughtful interpretation and

H. KAMERLINGH ONNES. „Further Experiments with Liquid Helium. G. On the Electrical Resistance of Pure Metals etc. VI. On the sudden change in the Rate at which the Resistance of Mercury Disappears."

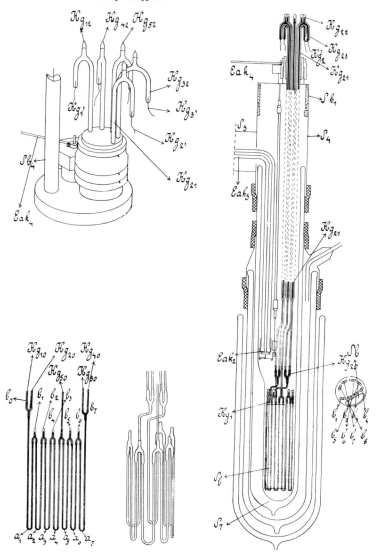

Figure 2.5   Sketch of one of Onnes' dewars. (Courtesy of the Kamerlingh Onnes Laboratorium, Leiden; copyright: Elsevier Science Publishers.)

analysis. Part of the problem, as pointed out by Casimir and others, was that liquid helium was scarce and available only on given days. Thus, researchers were then pressed to obtain as many data as possible during the 'run', that is, the time in which the liquid helium lasted in the apparatus. There wasn't much time to sit back and think, since the next run might not take place until far into the future!

3

The discovery of superconductivity

AFTER THE LIQUEFACTION OF HELIUM, Onnes and his collaborators spent a couple of years trying to reach lower and lower temperatures. Ultimately he (and his assistants) reached 1 K by pumping over the liquid helium bath, that is, using essentially the method described in the previous chapter. In 1910 he decided to abandon this project and started to concentrate on another keen interest of his, the study of matter at low temperatures.

3.1 The electrical resistivity of metals

Two concurrent interests were pushing men like Onnes and Dewar to look at the electrical resistivity of metals at low temperatures. One was thermometry; many experimenters realized that the variation of electrical resistance as a function of temperature could be used to make a thermometer out of a metal strip or wire, once a calibration had been carried out. (Such a trick is still in use today for certain applications.) Another method of measuring the temperature was to use a gas thermometer; in this case, Boyle's law (see previous chapter) was used to calculate the temperature from the measurement of the pressure and volume of a gas. However, gas thermometers were bulky and laborious to use.

Aside from Onnes's or Dewar's interests in thermometry, there were other fundamental reasons for investigating, at low temperatures, the behavior of the electrical resistance of metals as a function of temperature. Later we will examine more closely the mechanisms that were thought to be operative in the resistance to the passage of electrical currents in wires. Here it will suffice to say that there was a theory by Drude, later perfected by Lorentz, which proposed that the electrons which are responsible for carrying the electric charge in a metal could be thought of as behaving like a gas of particles. Many initially speculated (Onnes and Lord Kelvin among others) that at very low temperatures the electrical resistance would go up sharply since, near $T = 0$ K, the electron gas would freeze out and electrons would no longer be available for conduction. This wasn't an unreasonable idea; nowadays we can easily observe that water vapor becomes frozen when we open a freezer and observe that the food inside is covered with 'snowflakes'.

The purpose of Onnes's and others' experiments was to see if this bottoming out of the resistivity in metals and alloys was the prelude to a sharp rise. Contrary to these expectations, experimenters, including Onnes and Dewar, consistently found that for many metals the electrical resistivity dropped rather quickly with temperature down to solid hydrogen temperatures (around 16 K); at still lower temperatures Onnes found that the resistivity didn't drop as fast, but instead approached a very small value (see Figure 3.1). There was no evidence of 'freezing of electrons'.

Different samples were used but the result was almost invariably the same: there was no hint of a turn around in the resistivity as the temperature was lowered. Onnes and his collaborators studied different alloys of platinum and gold; from their measurements, they concluded that a very pure material would have a very low resistivity at the lowest temperatures. Onnes speculated that gold would reach zero resistivity at a temperature greater than 0 K. Then his attention turned to mercury, a metal that at room temperature is in the liquid state. He found that such a metal was easier to purify than most and he knew that impurities, even in small amounts, would have masked what he wanted to measure. For these reasons, he started to study the resistivity of mercury at low temperature.

His idea worked. At the Royal Society on April 28, 1911 he

reported that the resistivity of mercury dropped considerably at around 4.3 K (see Figure 3.2 – he later corrected this value when he calibrated his temperature scale). In the *Communications from the Physical Laboratory of the University of Leiden* he wrote:

... it was concluded that the resistance of pure mercury would be found much smaller at the boiling point of liquid helium [4.2 K] than at hydrogen temperatures [16–20 K] ... Experiment has completely confirmed this forecast.

But in his address at the ceremony awarding him the Nobel Prize in Physics he said:

as has been said, the experiment left no doubt that ... the resistance disappeared. At the same time something unexpected occurred. The disappearance did not take place gradually but abruptly [his underlining].

The *Communication from the Physical Laboratory of the University of Leiden* minutely yet tersely, described the procedures

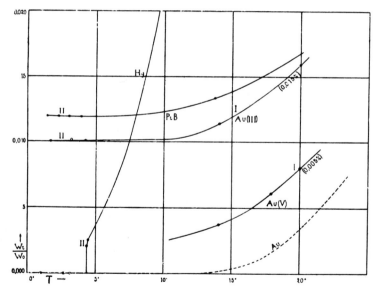

Figure 3.1   Resistance (vertical axis) versus temperature (horizontal axis) for different alloys. The dashed line is Onnes' extrapolation for a pure gold wire. Au stands for gold, Pt for platinum. Also plotted is the result for mercury (Hg) wires. (Courtesy of the Kamerlingh Onnes Laboratorium, Leiden. Copyright permission obtained from Elsevier Press.)

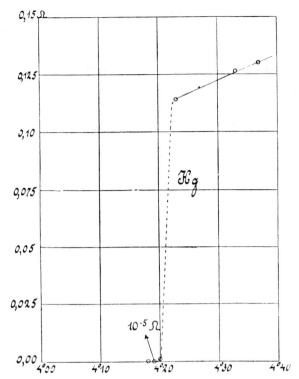

Figure 3.2   Resistance (vertical axis, in units of ohm) versus temperature (horizontal axis, in degrees Kelvin) for mercury (Hg) wires. (Courtesy of the Kamerlingh Onnes Laboratorium, Leiden. Copyright permission obtained from Elsevier Science Publishers.)

followed during the experiments and observations. But how different it was from his earlier writings on the liquefaction of helium in which his goals emerged from the report very clearly! Here Onnes had to confront observations which were entirely new, and it was not yet clear to him whether such a *drop* in resistivity could be explained as a complicated phenomenon occurring within an existing theoretical framework (as hinted in the first passage quoted) or as something *really new* that no theory known at that time could have predicted. The latter turned out to be the case.

While this observation was partially consistent with what Onnes expected, the sudden drop of resistance within the very narrow

Table 3.1   *Transition temperatures and critical magnetic fields*
(at 0 K) for selected elements.

Element	Critical temperature T_c (K)	Critical field H_{c0} (gauss)
Aluminum	1.12	110
Indium	3.41	280
Niobium	9.25	2060
Lead	7.20	800
Tin	3.72	310

range of a few tenths of a degree was puzzling indeed. Experiments
were repeated and the results were consistent with a vanishing
electrical resistance below 4.15 K. For all temperatures below
4.15 K that he was able to reach, the electrical resistance was zero.
When warmed up to above 4.15 K (to be called the transition or
critical temperature) the electrical resistance returned to its normal
values. Tin, another metal, was also found to display a similar
behavior with a transition temperature of 3.72 K, see Table 3.1.

It was in a *Communication from the Physical Laboratory* dated
1911 that Onnes first used the term supra-conductivity (or
superconductivity) to describe the passage of electric current in
mercury wires held at below 4.15 K with a resistance ten orders of
magnitude (or $1/10000000000$)† less than the resistance at 0 °C.

Later, Onnes built a superconductor in the shape of a ring and
injected a current into it. He then watched it flow undiminished for
24 hours; it would have taken a fraction of a second for an electrical
current to 'die' in an ordinary conductor. The physicist P. Eheren-
fest described this experiment to H. A. Lorentz in a letter dated
1914:

It is uncanny to see the influence of these 'permanent' currents on a
magnetic needle. [It was well known at that time that a current exerts

† Here we will introduce the following notation to cope with numbers which are either
very large or very small: 10^N means 1 followed by N zeros, such as: $10^5 = 100000$; 10^{-N}
means the reciprocal of 1 followed by N zeros: $10^{-5} = 1/100000$. 10^5 is pronounced: 10
to the power of 5 or, for short, '10 to the 5'. This notation is widely used by scientists
and engineers. Confronted with an ever increasing annual budget deficit and debt (in
the billions and trillions of dollars in the USA), politicians should consider adopting the
scientific notation.

a force on a magnetic needle.] You can feel almost tangibly how the ring of electrons in the wire turns around, around, around – slowly and almost without friction.

Nowadays we understand that superconductivity is a more complex phenomenon and that vanishing electrical resistance is but one of its most spectacular manifestations. But how did Onnes understand his own discovery?

Onnes believed that this drop in resistivity was associated with a quantum mechanical theory put forward by Max Planck and others (to be discussed in the next chapter). He didn't spend much time in theorizing since the experimental results were particularly unexpected and puzzling; for example, it was hard to explain how this transition could occur in impure as well as pure samples. Instead, he decided to concentrate on another set of important experiments.

3.2 Superconductivity and magnetism

Soon after the discovery of superconductivity, Onnes found that in superconducting materials electric currents could be propagated without resistance if they had less than a threshold current density, that is, current per unit area. If, for a conductor of a given size, a higher current was made to pass through, the conductor would revert to its 'ordinary or normal' state and exhibit an electrical resistance. This current is called the critical current. After many careful experiments, it was found that this was an effect associated with superconductivity and not with experimental artifacts; it was quantitatively but not qualitatively different for the various superconducting materials he investigated.

Onnes started to experiment with magnetic fields. Magnetic fields are important in operating everyday machinery as well as in research on the properties of matter. Practically everybody is aware of at least some properties of magnets. Magnetic fields can be generated by matter itself (as in permanent magnets, such as the ones used as paper holders on refrigerator doors) or artificially as in the case of electromagnets. An electromagnet can be made by winding an electric wire around a cylindrical support (see Figure 3.3). The magnetic field is proportional to the current passing through the wire and is directed, inside the cylinder, along the main axis; there is very little magnetic field present outside the coil. This is a very handy device, since one can generate a very uniform magnetic field inside the coil with little spill-over on the outside. However, there is a limit to how much current can be passed through wires, even if these are made of good electrical

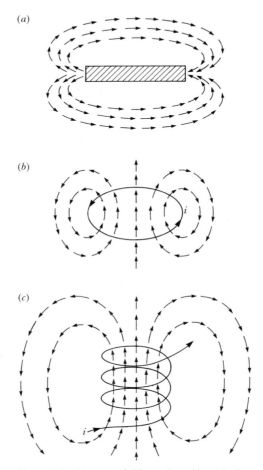

Figure 3.3   Magnetic fields produced by: (a) a bar magnet; (b) a current flowing in a loop of wire; and (c) a current flowing in a coil. Arrows indicate the direction of the magnetic field. Notice the similarities in the spatial distribution of the magnetic field in the three examples. In cases (b) and (c), the current i produces the magnetic field.

conductors such as copper, since the electrical resistance will cause a wire to dissipate energy in the form of heat (Joule heating). If this heat is not carried away fast enough, the temperature of the wire will become so high that the wire will melt or vaporize.

Measurements have shown that the electrical resistivity of metals goes down as the temperature goes down. Cooling the wires would seem to be a good method of reducing losses. Unfortunately, there is not enough difference in the value of the electrical resistance in copper, or other metals, at liquid helium temperatures and electrical resistance at room temperature to justify the cost of an expensive refrigerant such as liquid helium.

Such limitations, Onnes reasoned, wouldn't occur if the wires had no electrical resistance. Disappointingly, he and his collaborators found that magnetic fields exceeding a few hundred gauss (the strength of a small household magnet) destroyed super-conductivity. Nowadays this field is called the critical magnetic field and is commonly denoted by H_c. Onnes spent much of his time before World War I trying to overcome this limitation. His hope was to realize a 100 KG (1 KG = 1000 gauss) magnet. It took several decades to realize his dream.

3.3   Summary

Superconductivity, in its manifestation of zero electrical resistance, is a phenomenon which is easy for a low temperature practitioner to find and measure for the first time. We shouldn't forget, however, that it took Onnes many years of preparation to be able to observe such an unexpected phenomenon. He was the first to have the technical capability to achieve the low temperatures which were so critical for his discovery. In the end, the administrative skills, care for detail and acute observations of Heike Kamerlingh Onnes made the difference between success and failure. He had a clear vision of how to proceed and perceived that scientific discoveries depended upon specialized equipment and trained scientists; his was an approach far removed from the sometimes amateurish methods of nineteenth century science. Indeed, the essence of his methods is at the core of how scientific research is done today: one just cannot lock oneself up with a few assistants in a laboratory and expect to obtain news-breaking results. Instead, careful preparation, literature searches, consultation with other experts and design of new or improved instruments or methods lie at the heart of the most innovative discoveries today.

It is interesting to note how his discovery was made, that is, in the pursuit of purer substances and better thermometers. While the zero electrical resistance state is the most apparent manifestation of superconductivity, Onnes's work pointed to other curious relationships between superconductivity, electric currents and magnetic fields. Later we will see that one of the leads toward an understanding of superconductivity came from experiments with magnetic fields. There is an intimate connection between zero resistivity conductors, critical currents and critical magnetic fields.

It took another experimental breakthrough, twenty years after Onnes's discovery, to catalyze new efforts aimed at explaining these phenomena. This breakthrough, albeit less celebrated than Onnes's experiment, was achieved by Meissner and his graduate student Ochsenfeld. We will look at this new phase in the understanding of superconductivity after we have examined how some important physical phenomena were understood at the beginning of this century.

4

How electrical currents flow

ONCE THE TECHNOLOGY had been invented to do experiments at the low temperature at which liquid helium boils several laboratories in Europe started to work on superconductors. The most striking property of superconductors, the conduction of electricity with no resistance, was relatively easy to detect. But many questions remained unanswered. Why did certain materials become superconductors only at very low temperature? And why was it that others never became superconducting (at least at temperatures accessible in the laboratory)? Could metals, such as copper and silver which are good conductors of electricity, be made so pure that they offered no resistance to electrical current? Was superconductivity simply the state of a conductor when it was very pure? To understand the direction that Onnes's and others' research followed after the discovery of superconductivity in mercury and tin, we have to consider what those scientists knew about electric currents.

4.1 Electrons in wires

When we observe water running down a rain trough or a stream down a steep slope, we cannot but notice that the flow is not smooth and ordered, but rather chaotic, with many eddies and changes of path. Pebbles, twigs and other objects impede the smooth flow of water. Because of gravity, water eventually manages to overcome this resistance.

An electrical current is a flow of electric charges, in many ways analogous to the flow of water in a pipe. We can carry this analogy a step further. In both cases, there is a force that pushes the charges or the water along; there is also a flow measured as amount of charge flowing per second or cubic feet of water per second. Viscosity in water produces a resistance to the motion which eventually results in heat dissipitation. Is there an analogous dissipative mechanism in wires? Electric charges can flow more freely in certain materials, such as metals and alloys, than in others, such as plastics or stones. No matter what material is considered, with the exception of superconductors, as we will see later, some energy has to be spent to push these charges along. In other words, there is some resistance to the flow of the charges and a force, an electrical force, has to be supplied to overcome it; in the case of water, gravity or a pump supplies this force. The energy spent is expressed by the force times the length over which an object (water, charges) is pushed. Therefore, in the case of charges in a wire, the longer the wire, the more resistance it has (the electrical resistance is more precisely defined below). It is customary to write this energy, more properly called 'work', in terms of the potential difference, or voltage, which is the work to be done by the electrical force on a unit charge. Electric appliances are rated by voltage; small appliances work at 110 volts: others, such as dryers, at 240 volts. Electric appliances are also rated by how fast they can do a

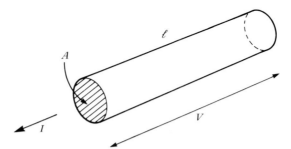

Figure 4.1   Section of a cylindrical conductor. A is the cross-sectional area, I is the current flowing and V is the voltage drop along the length ℓ.

certain amount of work. The work per unit time is called power and is measured in watts. For a given amount of current, the higher the resistance, the greater the amount of work that has to be supplied per second. An electric baseboard heater of 1500 watts will throw more heat into the room than one of 1000 watts.

Ohm empirically found many years before Onnes's discovery that the voltage applied across an electrical conductor, such as a piece of metal or the filament of a light bulb, is proportional to the current that goes through the conductor. The constant of proportionality, called resistance, depends on the type of substance the conductor is made of (copper – for a typical electric household extension cord; tungsten – for a light bulb) and its geometry. In short we write:

$$V = RI,$$

where I is the current (in amperes), V is the voltage difference (in volts) and R is the resistance in units of ohms.† For the electrical systems considered here, the power is given by the square of the voltage (that is, voltage times voltage) divided by the resistance. We can write it as:

$$W = V \times V/R \text{ or } W = V^2/R.$$

This power is dissipated as heat (as is the case for many appliances – an electric stove, for example) or as heat *and* light (as in a light bulb). For a cylindrical conductor we have:

$$R = \rho l/A,$$

where l is the length and A is the area of its cross section. (See Figure 4.1). The symbol ρ is the resistivity and gives the electrical characteristics of the material: it is the resistance for a wire of unit length and unit cross-sectional area; it is often more convenient to use than the resistance since it doesn't depend on the geometry of the wire. A very good conducting material, such as copper or silver, will have a small resistivity, of the order of 10^{-8} ohm × meter. A bad electric conductor (or insulator) such as glass will have a resistivity anywhere from 10^{10} to 10^{14} ohm × meter, that is, 10^{12} to 10^{16} (a thousand billion to ten million billion) times more

† The voltage drop in a wire with resistance of 1 ohm and current of 1 ampere is 1 volt. The power dissipated is 1 watt (see first formula). Household appliances consume anywhere from a few hundred watts (radio) to a couple of kilowatts (dryers, cookers).

resistance per unit length and area than an equivalent piece of copper! It is remarkable how this empirical law holds for such a wide variety of substances. Not every substance or device obeys Ohm's law; for example, diodes and transistors, present in consumer electronic items, don't.

Electric power is delivered from the power station to its users via strands of copper (or in certain cases aluminum) wires called cables. Notwithstanding the small resistivity of copper wires (which are several hundreds of miles in length in certain cases), appreciable power is wasted in heat irradiated away by the cables. Great excitement was aroused by the discovery of superconductivity; this is what Onnes wrote in 1911, the year of its discovery:

The possibility of using the super-conductors tin and lead gives a new departure to the idea of procuring a stronger magnetic field by the use of coils without iron ... even a coil of 25 cm diameter of lead wire, immersed in helium, could give a field of 100,000 gauss without perceptible heat being developed in the coil. Some such apparatus could be made at Leiden if a relatively modest financial support were obtained.

As we mentioned before, such developments didn't go very well, since even a modest magnetic field destroys superconductivity in elemental materials, such as lead, tin etc. Alloys, which are mixtures of two or more metals, have since been made, however, which have larger critical temperatures and can withstand much larger magnetic fields. Unfortunately the savings in power that would otherwise be wasted don't, even today, offset the expense of making special wires and installing special containers for producing and maintaining liquid helium. But if we had no resistivity whatsoever at room or even liquid nitrogen temperature, things *might* change. We will examine these materials and their commercial possibilities in a later chapter. Let us look now at how charges are moved through a wire and why heat is dissipated.

First, we should ask what the charges being moved through a wire are. Matter is composed of atoms, electrically neutral entities which make up ordinary macroscopic matter as we know it with our senses. Atoms contain an equal number of positive charges, or protons, and negative charges, called electrons. In reality the atom contains another particle which carries no charge, the neutron,

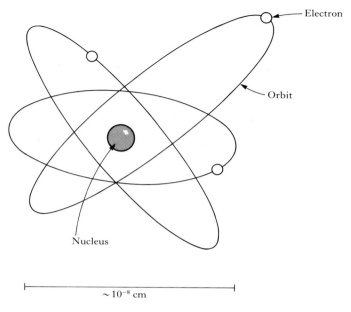

Figure 4.2   Schematic drawing of Rutherford's model of an atom. The nucleus with the positive charge is at the center. The negative charge is carried by electrons rotating around the nucleus. The diameter of the nucleus is about 10^{-13} cm, or 100000 times smaller than a typical electron's orbit.

which we need not consider here. Just about ten years before Onnes's discovery of superconductivity, J. J. Thomson showed that the electron, a particle with a mass $1/2000$ the mass of the proton, carries the negative charge. The charge carried by the electron is exactly opposite to the one carried by the proton. The current flowing in a light bulb is equivalent to about 10^{19} electrons per second (1 ampere).

How are the protons and the electrons arranged in an atom? At the beginning of this century there were various competing models. The one which emerged from the experimental work of Lord Rutherford and others had similarities to a planetary system (described in more detail in Chapter 6); electric charges of opposite sign attract each other, as masses of planets do. In passing we remark that there is only one type of mass and masses always attract each other; in the interaction between masses there isn't anything analogous to the electrical repulsion of unlike charges. In Rutherford's model the protons are concentrated in a central core, or nucleus, while the electrons move around it (Figure 4.2). If there is an attraction between a proton and the accompanying electron, why doesn't the electron fall into the proton? If they were at rest, they would indeed fall into each other, or rather, since the electron is so much lighter in mass, the electron would fall into the proton. If the electron has a velocity, the attractive force pushes the electron, at each instant, toward the proton, but because of the electrons inertia without complete success. The result, see Figure 4.3, is a circular orbit in which the electron 'falls' toward the proton at any segment of its path (called trajectory). A similar explanation holds for the notion of bodies which experience the gravitational pull, such as the Moon which constantly 'falls' toward the Earth. Don't the electron and the proton feel the gravitational attraction? Yes, they do, but it is very small compared to the electrical attractive force.

There are ninety-two types of atoms, or elements, each with one more electron (and proton, to maintain charge neutrality) than the preceding element.† Some examples are (with the number of electrons in parenthesis): hydrogen (1), copper (29), uranium (92). In order to explain how a current flows in a wire, we have to build

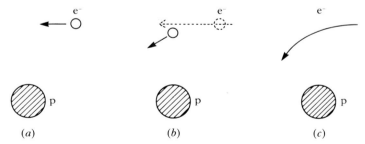

(a)   (b)   (c)

Figure 4.3   Interaction between a proton at rest (p) and an incoming electron (e⁻). (a) The electron is approaching the proton; (b) the electron is attracted toward the proton – the electron's trajectory starts to change; (c) the trajectory of the electron.

† Of course, there are millions and millions of different substances found in everyday life: water, alcohol, steel etc. However, they are all made by two or more of the ninety-two elements.

a model which describes the effect of the electrical resistance on the flow of electrons.

4.2   A model for the electrical resistivity

At the beginning of 1900, first Drude and then Lorentz put forward a theory to explain electrical conductivity in metals. In their view, only the outermost electrons of each atom of a metal (the ones with the trajectories very far away from the nucleus) are responsible for electrical conduction. In other words, when there is an electrical potential applied at the ends of a wire, only the electrons which are least bound to the nucleus of each atom are 'free' to move. The current is therefore produced by the movement of these electrons. Due to thermal agitation and in the absence of an external potential, the electrons move in all directions; such motion, on average, amounts to zero net displacement (see Figure 4.4). If a potential is present at the end of the wire, an electrical force will tend to move the electrons in one particular direction more than in any other; more precisely the electrons will go from the lower to the higher potential (or from the negative to the positive potential – see Figure 4.4). The resulting picture is electrons still moving in all possible directions but with a net

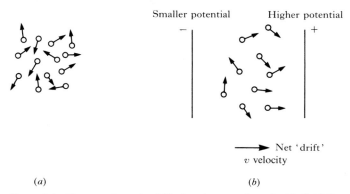

(a)   (b)

Figure 4.4   Electrons in a wire. (a) In the absence of an electric field, the motion of the electrons is random. The sum of the velocities of all the electrons is zero. (b) When an electric field is applied, a difference in electrical potential appears across the wire. The sum of all velocities of the electrons is not zero but directed as shown. The net movement of the electrons is toward higher electric potential (sign ' + ').

velocity in the direction of the applied electrical force. The resistivity can be obtained by working out the results of the collision of these electrons with the ions (that is, the atoms minus the free electrons) of the wire. While this model has been proven wrong because it doesn't account for quantum effects, to be discussed below, the main physical picture exploited in the model has turned out to be correct. In fact, this model, however crude, was remarkably successful in explaining many observations.

We call this model a classical one, that is, one that used classical, Newtonian laws and not quantum mechanical ones. To appreciate this distinction we have to recall briefly what the understanding of physical phenomena was at the end of the past century.

It was a happy end-of-the-century insofar as physics was concerned, or at least this was the prevalent opinion of many experts in the field. James Clerk Maxwell had just shown how electricity and magnetism were intimately related; Newtonian mechanics (which regulates the motion of macroscopic objects, from a baseball to a star) was very well known and had been successfully applied to many problems; thermodynamics and optics were also quite well understood. Yet, there were a few puzzling discrepancies here and there between some experimental results and accepted theories, but the general view was that these differences could be reconciled by doing more sophisticated experiments and calculations.

However, the pace of events started to change rapidly at the turn of the century. In 1900 Max Planck, in order to explain some experimental results concerning the emission of light by bodies which couldn't be accounted for by current theories, proposed that atoms vibrate around their equilibrium positions as harmonic oscillators, such as small balls do when they are attached to a spring and set in motion. However, and here lies the revolutionary aspect of his model, the energy of these oscillators was quantized, i.e., only certain values of energy (and, as he showed, of frequency!) were allowed according to:

$$E = (n + \tfrac{1}{2}) h\nu,$$

where h is a universal constant (called Planck's constant) and ν is the frequency of vibration of the oscillators; n is a positive integer (see Figure 4.5). Thus, the energy of the oscillator can change only

Energy

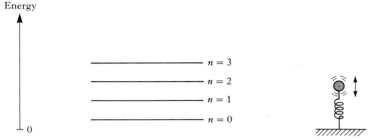

Figure 4.5   An oscillator is here sketched as a ball attached to a spring. According to Planck, the oscillator can have only certain energy values. The first four energy levels are shown. The higher the 'n' number, the higher the energy and the frequency with which the oscillator vibrates.

by a finite (discrete) amount called a 'quantum'. Notice that for the minimum value of n, $n = 0$, there is still some energy left in the oscillator. Even at 0 K, these atoms would still oscillate! This was indeed surprising and even upsetting to many physicists. At the time of Onnes's discovery there was no theory which used these quanta in calculating the electrical resistance. However, Planck's work had an influence on the work of Onnes, as we will see shortly. One might ask: if what Planck said is true, how is it possible that nobody detected or needed quantized energies to explain most of the observed phenomena? The answer lies in the fact that these quantum effects are important at the atomic scale; in a few cases they have a way of showing up in macroscopic phenomena and superconductivity is one of these. Otherwise, quantum phenomena are quite inaccessible except in very specially designed experiments.

Using quantum mechanics, a theory developed in the twenties which used quanta of energy to explain properties of the atomic world, we can now calculate how electrons move in a metal and what is the influence of impurities, disorder and vibrations of atoms from their rest position due to temperature. We now know that the vibrations of atoms at 0 K will not disturb the motion of electrons in an ideal, perfect crystal, but impurities or vibrations at temperatures higher than 0 K will. However, these theories were developed much later than Onnes' discoveries; thus he had to try to fit his observations into what was known at that time about electrical conductivity at low temperatures.

4.3   Onnes's plan

'What happens,' Onnes asked, 'to the resistivity if we reduce the temperature of a wire to near absolute zero?' If you believed, as many did, that the 'classical' (as opposed to quantum mechanical) picture of physical phenomena was satisfactory, you would not expect anything startling to be discovered. Because ions of the solid (atoms which have acquired or lost one or more electrons in their outer orbits) would have less energy, fewer collisions with electrons would occur and the resistivity would go down as the temperature approached o K.

On the other hand, if you took the hypothesis of Planck seriously, you could accept the idea that something remarkable might happen at low temperature! This was the question that Onnes, Dewar and others wanted to answer. There were a few different models or conjectures from which one could start working. Matthiesen discovered (in 1864) that the resistivity didn't go to zero as T was lowered (see Figure 4.6) but seemed to approach a residual value which depended on the impurity content (the higher the concentration of impurities, the higher the residual resistivity; different kinds of impurities might have quantitatively different effects). Dewar thought that the resistivity would go to zero as the temperature went to zero, since at smaller and smaller tempera-

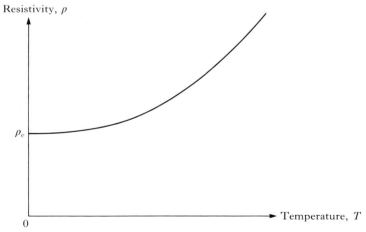

Figure 4.6   Resistivity ρ as a function of temperature (T) near 0 K for a metal or alloy.

tures the atoms would have fewer and fewer vibration amplitudes which could interfere with the electrons' movements. On the other hand Lord Kelvin, using the analogy that the mobile electrons in a metal form a kind of 'electron gas', thought that the resistivity should go up due to the condensation and freezing of the gas of electrons on the very atoms to which they belonged. In other words, the electrons wouldn't be able to move when pushed by an electric force and no current would flow.

Onnes's thoughts were influenced by the work of Planck. He concluded that *impurities* would freeze and electrons would be able to move around the crystal freely without bumping into atoms. Therefore the resistivity would go to zero with the temperature. He then set up his apparatus to measure the electrical resistivity of metals and alloys. Some of his data suggested that if the level of impurities was reduced by carefully choosing very pure materials, the resistivity would become very small at a temperature above 0 K. In 1910 he concluded that this was true in the case of gold and platinum wires (Figure 3.1). However, his apparatus to measure electric resistance wasn't sensitive enough to prove his point unequivocally and his samples still had impurities. He then decided to use mercury wires because he could prepare these with a high degree of purity.

In light of the discussion above, the fact that the resistivity of mercury dropped to zero (or, more precisely, to an undetectable limit for his instruments) was not very surprising. His writings didn't even assign much importance to the discovery. To find out how low the resistivity went, he spent considerable time perfecting his instruments to measure the resistance. Yet, he always found that the resistance dropped below the detectable limits of his best instruments. The real surprise was another one: his instruments always showed that the resistance dropped *abruptly*, see Figure 3.2. He was inclined to think that superconductivity was a quantum phenomenon, but, as we will see, neither he nor others were able to make much progress in explaining these results. Many more years had to pass before a successful explanation was proposed for these and many other puzzling experimental results.

What did Onnes accomplish? As we said earlier, a few physicists at that time could have come across, perhaps by chance, the zero resistance behavior as discovered by Onnes, if liquid helium had been more readily available. Indeed, the difference between a good and a mediocre scientist often lies not in his ability to make the discovery, since this can occur by chance, but in the ability to recognize and exploit it. Onnes wasn't just an ordinary scientist who by chance measured the resistance of mercury at low temperature (his preparation and attention to detail attest this). Furthermore, he didn't spend much time in theorizing; instead, he and his assistants went back to the laboratory and busied themselves with improving the apparatus and performing other experiments with magnetic fields and other materials. This work didn't solve the puzzle of what was causing superconductivity, but eventually led other experimental groups and theoreticians down the right path. We will see how in the next chapter.

4.4   Is a perfect conductor a superconductor?

To summarize, our present understanding of how electric currents flow in ordinary matter tells us that an electrical current is impeded in its flow by impurities and defects in the material and by the disordered motion of atoms or ions in the wire. If one could make very pure materials and hold them at 0 K to suppress the disordered motion of atoms, one could obtain a perfect conductor (at 0 K); but in reality, a superconductor is not just a perfect conductor. The onset of superconductivity can be observed in the materials mentioned above at temperatures which are clearly higher, although by not much, than 0 K; furthermore, super-conductivity also exists in materials which are quite dirty (that is, with significant concentrations of impurities) and in which the arrays of atoms are not as perfectly organized as in single crystals. What we know about conduction of charges in ordinary matter cannot explain this behaviour. Thus, there must be something else that characterizes superconductivity. But what? Onnes came very close to finding out, even though he didn't realize that some of the experiments he was doing held the key to understanding super-conductivity. To see how this happened we have to turn to magnetism.

5

A breakthrough: The Meissner effect

5.1 Research resumes after the War

AFTER THE WAR, research on superconductivity resumed in earnest. Several laboratories in Europe were very active in low temperature physics. An especially important development was the increased availability of helium gas (due to the supplies coming from America, where helium was more abundant), necessary to produce the liquid helium essential in lowering the temperatures to the level in which superconductivity phenomena appeared.

For many scientists this little known gas was far more precious than a layman could appreciate.

Kurt Mendelssohn recalls that:

The arrival of the first American helium from gas wells was an important event, and I went down to the Berlin Customs House to open the crate. But then the official wanted to open the crate too, to see whether it contained liquor. He only relented when I assured him that I should be most disappointed if it did.

As the deliveries of helium resumed, the research program on superconductivity was restarted at Leiden. There was a great deal of work to do. Early attempts by theorists to explain Onnes's 1911 discovery were conspicuously unsuccessful (even Albert Einstein tried!). The decision of Onnes and others at Leiden and in other parts of Europe (notably Berlin and Oxford) to continue experiments aimed at characterizing superconductors was a wise one. In fact, by comparing various physical properties of superconductors with those of normal metals (here defined as the ones which fail to become superconducting at any experimentally accessible temperature), they hoped to find the key that differentiated the two. With luck, and through pondering these differences, theorists were able to come up with some explanations as to what caused these baffling properties.

Many more elements and alloys were found to be superconductors, all of them with transition temperatures near the liquid helium boiling point, that is, slightly above 4 K. Another characteristic of these materials was the instability of their superconducting state when placed in an external magnetic field. If a magnetic field of a few hundred gauss was applied to a substance in its superconducting state (for example, a magnetic field produced by a typical household magnet), that substance would revert back to its normal (metallic) state, thereby losing all its superconducting properties. Onnes and his collaborators measured, for each temperature, the minimum magnetic field that destroyed superconductivity (called the critical magnetic field). They found the trend shown in Figure 5.1. The region below the curve represents values for T and H for which superconductivity is present. For values of the magnetic field above the line $H_c(T)$ in Figure 5.1, the metal goes back to its normal state.

The discovery that modest magnetic fields could destroy superconducting properties in metals was disheartening. Even worse, first Onnes and his collaborators, and afterward many other groups in England and Berlin, found that it was not possible to send large electrical currents through the superconducting wire; rather there was, for every material with superconducting properties, a maximum current which could be sustained. It was later understood that these two phenomena are connected, since an electrical current generates a magnetic field which can contribute, together with other effects to be discussed later, to the destruction of superconductivity if this field is large enough. Obviously, these discoveries were quite disappointing because the superconducting materials known at that time couldn't be used for many of the practical applications they had been thought to be ideal for, such as

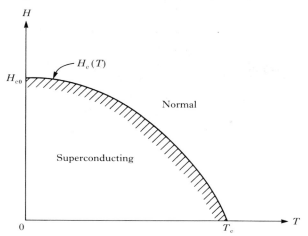

Figure 5.1   Critical magnetic field curve $H_c(T)$ versus temperature (T). Above the curve, the material is in its 'normal' state. Below it, the material is in the 'superconducting' state. This type of curve defines the so-called 'Type I' superconductors. H_{c0} is the value of the critical magnetic field at 0 K.

the generation of powerful magnetic fields by circulating electrical currents in windings of a solenoid (or coil) or the transmission of electrical power over large distances without energy loss.

5.2   A hollow versus a solid sphere

Let us go back to the experiments in Onnes's laboratory. Onnes and Tuyn perfected a method of measuring residual electrical resistance, if any, in the superconducting state. They put a ring of a superconducting metal in a magnetic field at a temperature high enough to maintain the metal in its normal state. They then cooled the ring below the superconducting transition temperature and switched off the magnetic field (see Figure 5.2). It was well understood in those days that a change in the magnetic field with time produces an electrical current in a ring located within the magnetic field. This current is higher if the rate at which the magnetic field changes is faster. This law of electromagnetism, the law of induction, was discovered by Faraday and then reformulated by Maxwell in the nineteenth century. The principle is the same as the one which is exploited in the ignition coil of a car, whereby a high voltage (necessary to generate the sparks) is produced with the aid of the change of a magnetic field and the law of induction. In an ordinary metal, this induced electrical current lasts only a very short period of time, because, once the magnetic field has ceased to change, the current decays due to the resistance it encounters. This effect produces heat (called Joule heating) and, in the absence of an external force to push the electrons along, the current will quickly come to a stop. Obviously, in a superconducting material, this

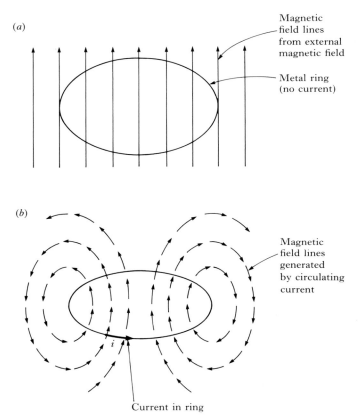

Figure 5.2   A wire in the shape of a ring is placed in an external magnetic field (a). The ring is cooled below the superconducting transition temperature and the external magnetic field is switched off. Because of Faraday's law, a current circulates in the ring (b) to offset the change in the magnetic flux. If the ring is in the superconducting state, a steady electrical current will circulate. If the ring is in the normal state, the current will die out very quickly.

current would go on 'forever', since in this case there is no electrical resistance.

Tuyn and Onnes set up the induced current in a ring and then watched how much it decreased as a function of time. For the several hours during which they could hold the liquid helium in the dewar, the current changed very little; they were able to set unprecedented limits for the accuracy with which they claimed that a superconducting sample had zero electrical resistance. Nowadays, similar experiments have been carried out for a period of a year or more.

Later researchers repeated the ring experiment with a hollow lead sphere and obtained comparable results. Unfortunately, the fact that they did their experiment with a *hollow* instead of a solid lead sphere was forgotten or misread; this mistake hindered progress in understanding superconductivity for several years. Before we see why it was so crucial to do an experiment with a solid sphere, let us see what other experiments were being carried out in low temperature laboratories throughout Europe.

Experiments were conducted to find out whether the change in electrical characteristics of a metallic sample (from a normal to a superconducting state) was followed by other transformations, such as a change in the lattice constant (that is, the characteristic distance between atoms in a solid). Keesom, who took over the work of Onnes after his retirement, found that there was a jump in heat capacity at the critical temperature but no latent heat. The latent heat is the amount of heat one has to give to or absorb from a sample to complete the transformation of a substance from one phase to another, for example from the liquid to the vapor state of water; the heat capacity is the amount of heat that a substance must absorb to change its temperature by 1 K. This observation was important, because phase transformations without latent heat exhibit different characteristics than those that have it and the former are more easily described by theoretical models. Moreover, while the determination of the thermodynamic behavior (temperature, heat capacity, pressure etc.) of a material cannot give us a true understanding of the microscopic mechanisms responsible for certain phenomena, it can organize our thoughts and point toward useful connections between different physical processes, as we will see shortly.

The interests of the experimentalists therefore shifted from the measurement of electrical properties to thermodynamic ones (such as heat capacity, latent heat, etc.). However, there were several experiments and considerations which pointed away from an explanation of superconductivity based on thermodynamic arguments. Here we will mention two arguments which influenced the field up to the next most decisive experimental discovery in 1933.

Firstly, the change of electrical resistivity in the metal from its normal value to zero (as in a superconductor) was perceived by many as a radical change in a microscopic property of the substance, and not as a change in the thermodynamic state of the substance, as in the change from vapor to gas. The most accepted 'explanation' was that the mean free path of the electrons, or the distance traveled between two subsequent collisions, increased dramatically and abruptly at the critical temperature. If the mean free path was very large, only a very small electrical resistance would be observed in the passage of a current since the electron would collide less frequently. The reason for this postulated change in the mean free path of the electrons was, however, wholly unknown.

Secondly, the possibility of setting up electrical currents with no dissipation (a direct and undisputed consequence of zero electrical resistance) implied thermodynamic irreversibility. Therefore, a description of superconductivity in terms of ordinary equilibrium (that is, reversible) thermodynamics would be useless. To see this, let us consider the following experiment. Suppose that we have a solid (!) lead sphere in a magnetic field at a temperature in which lead has normal electrical resistance. The magnetic field will penetrate inside the lead sphere (see Figure 5.3(a), position A) as it does most materials except ferromagnetic ones (more commonly known as 'permanent magnets'). Let us cool the sphere below the transition normal-to-superconducting point (position B in Figure 5.3(a)) and then remove the magnetic field (position C). When we remove the magnetic field, electrical currents circulate in the sphere because they are induced by the rate of change of the magnetic field (the above-mentioned Faraday's law) of induction. Since the sphere is superconducting, these currents will circulate virtually forever. These are called supercurrents. But it is also well known that electrical currents produce a magnetic field of their

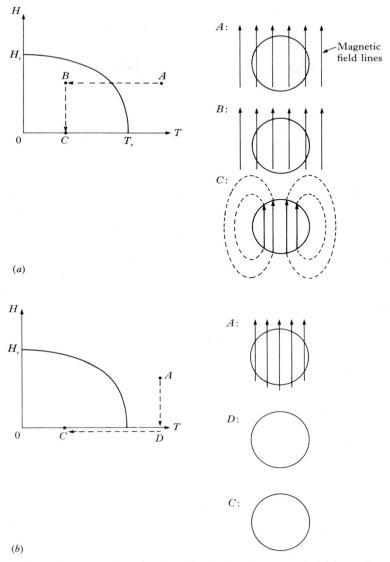

Figure 5.3(a) Experiment of cooling a lead sphere in a magnetic field assuming there is no Meissner effect. When the magnetic field is switched off (from B to C), currents in the sphere prevent this change (Faraday's law) by producing a magnetic field of their own. The lead sphere remains magnetized at C. (b) When the magnetic field is switched off (from A to D), an electric current is set up on the sphere (Faraday's law) but it decays quickly because of the electrical resistance of the sphere. The sphere in D and C is not magnetized.

own. Therefore, and this can be shown precisely but will not be reported here because of technical complexity, the induced electric currents will maintain the magnetization inside the sphere even after we have switched off the magnetic field. The final state of the sphere is the one with a magnetic field inside.

Let us consider the following alternative experiment (Figure 5.3(b)). Starting from A we turn off the magnetic field. At D the induced current dies down quickly because of the resistivity of lead. Thus at D there is no magnetic field in the sphere. We cool the sphere through the critical temperature but without bathing it in a magnetic field. We reach the same temperature for the sphere as we had in the experiment above (compare Figure 5.3(a) and 5.3(b), position C); now, however, there is no magnetic field either inside or outside the sphere. Thermodynamically, the final state is the same $(T = T_0, H = 0)$, but we have obtained two physically different spheres one with and the other without a magnetic field depending on the path we followed. It is a basic tenet of equilibrium thermodynamics that the state of a substance can be described by macroscopic variables (such as pressure, volume, magnetization, temperature, etc.) and that the state (called the thermodynamic equilibrium state) should not depend on the particular transformation (such as a change in temperature or pressure) or path which was followed to go from one set of values to another. Of course there are many examples in which this does not occur; a piece of ordinary glass, for example, is a substance which has been cooled too rapidly and didn't have the chance to rearrange its own atoms in such a way as to be always in thermal equilibrium with itself or the outside. In this case we say that we have achieved an irreversible transformation. If we follow the same path but backwards and now heat the substance, it will generally not go back to the original state. Irreversible transformations are possible in nature and occur all the time; they are, however, much harder to understand and describe. It was entirely possible, although disappointingly so, that these transformations with superconductors were irreversible. In that case, theoreticians would face an uphill battle to come up with explanations for these phenomena.

To summarize, in the late twenties superconductivity was still unexplained, although a much larger amount of data than fifteen years before was available. On the one hand, certain properties pointed to a thermodynamic transition between a normal and a superconducting state; on the other hand, there were measurements which showed clear signs of thermodynamic irreversibility. To clarify the influence of a magnetic field on the properties of a superconductor, Walther Meissner and his graduate student Robert Ochsenfeld (in Berlin) initiated a well planned research program.

5.3 Meissner's discovery

Meissner and Ochsenfeld wanted to know whether the transition from normal resistivity to superconductivity was connected to a change in the magnetic properties of the material, such as a change in its magnetic permeability or propensity to become magnetized. They also wanted to find out whether supercurrents, circulating on the skin of the conductor, shielded the magnetic field completely and prevented it from penetrating inside the superconductor. They repeated the experiment described in Figure 5.3 but this time with a solid cylinder. But what was started as an almost routine experiment gave an entirely unexpected result! What they found was that the magnetic field was *expelled* from the interior of the sample (see Figure 1.2); the ability to (partially) expel a magnetic field is called diamagnetism; in the case of superconductors, the expulsion is total and no magnetic field is left inside. Another way of saying this is that the sample is perfectly diamagnetic. All materials, whether superconducting or not, exhibit diamagnetism although not to the same degree. In most materials, this effect is typically tiny, many times smaller than other magnetic phenomena; consequently, the magnetic field is barely diminished, let alone expelled, from the interior. What Meissner and Ochsenfeld saw was a *large* and unexpected effect. Onnes's unexpected and clear results of perfect conductivity (vs perfect expulsion of the magnetic field) come to mind.

If we take this effect into consideration, is the outcome of the experiment described in Figure 5.3 changed? Let us repeat the experiment described in Figure 5.3, but with knowledge of both Onnes's and Meissner's experiments. The main difference between the experiment in Figure 5.3 and the new one, see Figure 5.4(a), is that in going from position A to position B, the magnetic field is expelled from the sphere as it becomes superconducting. When the

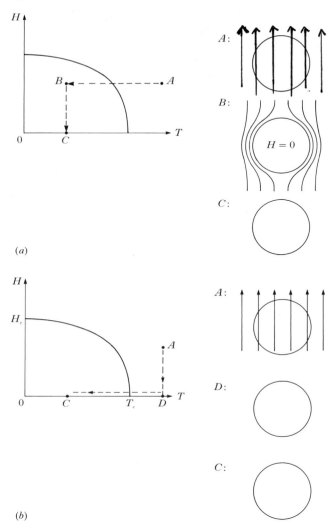

Figure 5.4(a)   Experiment of cooling a lead sphere in a magnetic field assuming the Meissner effect. When the sphere becomes superconducting (from A to B) the magnetic field is expelled. The sphere is not magnetized when it is taken to C. In B magnetic field lines are denser (that is, closer to each other) to indicate that the magnetic field just outside the sphere is stronger. (b) When the indicated path is followed, the sphere reaches C in a non-magnetized state. This state is the same as the one reached using the A to B to C path of (a).

external field is switched off, no magnetic field remains inside, (position C). Thus, we can get to position C following the two different paths (compare Figures 5.4(a) and (b)) and end up with the same physical system of no magnetic field inside the sphere. We say that we have achieved thermodynamic reversibility.

Now we are ready to answer the question: what difference does it make if we use a *hollow sphere rather than a solid one*? Sometimes what is perceived to be just a small unimportant detail puts people on the wrong track! Remember that quite different results were obtained by employing a hollow versus a solid sphere. Now that we know of Meissner's discovery let us repeat the previous experiment with a hollowed sphere and let us suppose that the walls are very thin compared to its radius. If such a sphere is placed in a magnetic field when in its normal state and then cooled so as to become superconducting, there is no magnetic field to be expelled from its interior, therefore no electric currents are set up (we will discount the slight contributions from the metal within the thin walls). If the magnetic field is then switched off, an electrical current is set up, or 'induced', around the sphere to prevent such a change (the Faraday effect discussed above). At the end of this process we have a sphere in the superconducting state which has an electric current flowing within its thin walls and around its perimeter. A different situation is reached if the sphere is cooled without a magnetic field. Then the sphere will be superconducting but will *contain no* circulating currents since the Faraday effect hasn't intervened in this case. Depending on how the experiment is carried out we get two different answers, even if its thermodynamic status, as described by the temperature, for example, was the same.

At first glance, Onnes's experiment supported the irreversibility hypothesis; however, reversible processes do occur (as demonstrated by the solid sphere experiment) though in the case of the hollow sphere they are masked by other effects (due to the walls, for example). In reality, more elaborate and subtler explanations are necessary, but their complexity prevents greater elaboration here.

Before Meissner's discovery we had to explain perfect conductivity (Onnes's experiment in 1911). Now we have to explain Onnes's perfect conductivity and Meissner's perfect diamagnetism. Surely, we are worse off than before! Now we have to explain how a superconductor can behave as a perfect electrical conductor *and* a perfect diamagnet.

Strange as it may seem, however, it has become easier to explain superconductivity. Here is how. Superconductivity (perfect conductivity and Meissner' effect) can now be described as a thermodynamic state. The passage from the normal to the superconducting state is a phase transformation of the electronic structure with a characteristic change in the specific heat. It was also possible to have a plausible explanation of the destruction of the superconducting state by a sufficiently strong magnetic field. The superconductors, or rather the supercurrents, do a certain amount of work to shield out an external magnetic field. When this work is not compensated by a reduction in energy due to the absence of magnetic field inside the conductor, there is no incentive for the magnetic field to stay outside the conductor and, in fact, the magnetic field starts to penetrate the metal.

John Bardeen, twice a Nobel Laureate in Physics who, in 1957, was one of the formulators of the most successful and complete theory of superconductivity that we have, said:

While an adequate theory must explain both aspects, the diamagnetic approach has been the most fruitful in indicating the nature of the superconducting state.

This connection with thermodynamics was important because it organized some of the various experimental observations and spurred productive theoretical work, as we will see shortly. On the other hand, while thermodynamics gave a useful description and organization of the various observed phenomena, it didn't furnish any explanation of superconductivity at the atomic level. For example, there were no theories which predicted whether a given material would become a superconductor and at what temperature, or how electrons organized themselves to produce the observed phenomena. To see how a microscopic explanation finally emerged, we have to know a little bit more about what theorists did in the meantime.

5.4 Theorists look at experiments for inspiration

In the years between the two World Wars many important advances were made in superconductivity. Numerous experiments were carried out and several points were clarified. Nonetheless, a comprehensive understanding of all the phenomena associated with superconductivity was still missing. To theoretical physicists the prospects of formulating a theory of superconductivity seemed remote – remote enough to prompt Felix Bloch, a well known physicist who at that time worked on the theory of matter in solids, to say, jokingly:

The only theorem about superconductivity which can be proved is that any theory of superconductivity is refutable.

Meissner's discovery, while much less publicized than Onnes', proved to be just as seminal. Several researchers now attacked the problem in different parts of Europe. In Holland, Gorter and Casimir worked on the magnetic properties of superconductors. They proposed that the electrons, which carry the charge in a metal, come in two flavors: normal and superconducting. The first behave as electrons do in a normal metal (that is, they obey Ohm's law) while the others are responsible for the superconducting properties. As the temperature is lowered beneath the critical temperature, more electrons become the superconducting type and fewer remain the normal type, in such a manner that the total number of electrons is still the same. While there was no direct experimental evidence for the existence of these two types of electrons, their model, or two-fluid model as it was called, explained a few puzzling experimental facts and, above all, proved to be very fruitful in the years to come.

In addition to Holland and Germany, England became an active center, with the work of Mendelssohn on superconductivity in alloys and the arrival from Germany of the London brothers, Heinz and Fritz. As is well known, those were difficult years in Europe. Several scientists fled Germany, while others (Casimir, for example) traveled to attend international meeting in Germany only after making painful decisions. Still others, such as the scientist Shubnikow from Russia, never made it to any of these international conferences (such as the 6th International Conference on Low Temperature Physics in The Hague in 1936; Shubnikow was executed a few months afterward).

Mendelssohn summarized well the perils of working in that period:

As I was leaving the University library after reading Keesom's paper, I had to duck a few bullets which were flying through the streets of Bresleau, heralding the approach of the Nazi rule, and it became clear that the cooling experiment had to be deferred a little. In fact, it was eventually carried out at Oxford in 1934.

The Londons worked on the problem of the relationship between the magnetic field and electrical currents in super-conductors. As mentioned previously, a circulating electrical current produces a magnetic field. Meissner showed that when a metal becomes a superconductor, the magnetic field is expelled from the interior. The 'Meissner effect' can be explained by reasoning that electrical currents set up on the surface of the superconductor produce a magnetic field equal in strength but opposite in direction to the external one, thus making the magnetic field inside exactly zero. The effectiveness of these currents (or supercurrents) which act as a shield against the magnetic field is

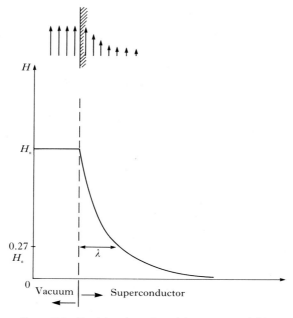

Figure 5.5   Partial penetration of the magnetic field at the surface of a superconductor. Significant penetration occurs over a distance λ called the London penetration depth.

not perfect, and in a small region near the surface there is some penetration of the magnetic field. The Londons' theory predicted how the magnetic field died out as it penetrated the conductor; a characteristic length (λ) over which the magnetic field decreases to 27% of the original value was given in terms of the mass and charge of the electron and the density (number of electrons in the sample divided by the volume of the sample) of 'superconducting electrons' (or electrons which were thought to be responsible for superconductivity), see Figure 5.5. We will encounter this characteristic length later when we will illustrate different types of superconductors.

5.5   What's missing?

These events took place in a restricted period of time, from 1932 to 1936. In hindsight we can say that by that time most of the pieces of the puzzle were on the table. Researchers knew about perfect conductivity and diamagnetism, magnetic and thermodynamic properties of superconductors, and many other phenomena related to superconductivity. There were also a few theories which attempted to describe, rather than explain, these phenomena. These theories were inspired by experimental findings or phenomena and are commonly called phenomenological. They help organize the facts we know in a more coherent fashion and sometimes lead people to make productive connections between different phenomena thought to be unrelated. On the other hand, they cannot give detailed reasons for why things happen a certain way based on the known laws of physics, nor are they capable of making predictions about phenomena not yet observed. The theories of the Londons and Gorter and Casimir are examples of phenomenological theories or models. However, notwithstanding the work of many brilliant minds, a theory that explained superconductivity using the constituents of superconductors, that is, electrons and atoms, was still lacking.

Help came from another direction. In those years (in the twenties and thirties), a revolution was taking place in physics: the well established world of what we now call 'classical mechanics', which is essentially based on the well known laws of Newton, was supplemented by 'quantum mechanics'. We have already seen briefly how the idea of quanta, or discrete amounts of energy came about. We must now learn a few things about how theorists used this idea to explain many more phenomena of the atomic world. In fact, quantum mechanics holds the key to an explanation of superconductivity.

6

Quantum mechanics and superconductivity

6.1   The strange world of quantum mechanics

FOR CENTURIES, the learned men of the ancient world tried to fit natural phenomena into their own preferred philosophical theories. Phenomena which didn't fit these schemes were either dismissed or provided with special, *ad hoc* 'explanations'. Galileo Galilei and Isaac Newton are now recognized as the fathers of the scientific method because they let experimental observations guide the formulation of their theories, instead of trying to fit data in a scheme of thought based on philosophical appeal. Isaac Newton derived his famous three laws of motion after experimental observations became available to him. These observations covered the world of Nature which was accessible to human observation and interrogation, from ordinary objects on Earth to celestial objects bright enough to be detected by the crude instruments of that time. Because no probe had yet been invented to look at the atomic and subatomic world, Newtonian mechanics had not been tested on objects approaching the size of an atom (1×10^{-8} centimeters or 5 billionths of an inch).

In what is now known as 'classical mechanics' (in contrast to 'quantum mechanics' described below), the motion of an object can be predicted by solving Newton's second law, the famous

$F = ma$ where F is the force, a is the acceleration (the rate of change of the velocity) and m is the mass of the particle. By (physical) law we mean here a set of rules about certain quantities or variables; it is usually written in mathematical form called an equation as shown above. In this case, once the force is known the velocity can be predicted as a function of time. We then say that we have solved the equations of motion.

The beauty of these laws is that they allow us to predict with certitude how an object will move, from a baseball bat to the Moon revolving around the Earth. The 'only' thing we have to do is specify the forces acting on the object we are studying and the position and velocity of this object at a given instant of time. Using mathematical formulas and other 'tricks' we can calculate where the particle will be (or has been) at any other instant. These laws work so marvellously that we can predict the relative positions of planets and comets years and years in advance. Newton's laws for the motion of bodies and Maxwell's laws for electromagnetic phenomena were able to explain most of the observed physical phenomena at the end of the last century. At that time, more than one scientist believed that our knowledge of the physical world was adequate. Further improvements were possible, but the Newtonian–Maxwellian construction seemed on solid ground. In fact, a world of change was going to transform physics in a very short period of time.

Lord Rutherford was a big figure (and not only figuratively! – see Figure 6.1) in European scientific circles in the early part of this century. His careful measurement of the collisions of atomic particles (helium nuclei) with metal targets gave us one of the first glimpses into the atomic world. From his experiments Rutherford deduced that an atom consisted of a nucleus, where the positive charges were located (protons), and electrons whirling around them as in a miniature planetary system with its sun (nucleus) and its planets (electrons). The model, however, had its shortcomings, although the main picture was essentially correct. The major problem with this model was that, according to the laws of electromagnetism, an electron going around a nucleus would lose energy because it would emit electromagnetic radiation. If an electron loses some of its energy, its velocity starts to decrease. The electron then gets closer and closer to the nucleus until it falls into

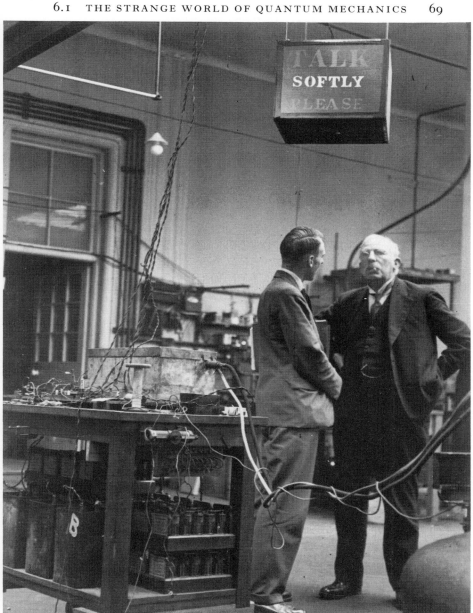

Figure 6.1   Ernest Rutherford (right), ca. 1935. (Courtesy of the AIP Niels Bohr Library.)

it. Readers should be reassured that despite the analogy of an electron with a planet (and of electrical with gravitational forces), the Earth will not fall into the Sun any time soon, since it doesn't emit radiation as an electron (or any accelerated charged particle) does. Despite what the laws of electromagnetism and Newtonian motion predicted, electrons didn't fall into the nucleus. Obviously something was wrong with the theory. But how was it possible to fix it?

A decade later Niels Bohr, a Danish physicist, took the model a step further by incorporating Planck's quantization. As we have seen previously, Max Planck, at the beginning of this century, was the first to introduce the quantization of energy. According to his hypothesis, energy is exchanged in discrete amounts. Bohr's model required that electrons move in quantized (that is, discretely located) orbits around the nucleus, see Figure 6.2. An electron moving in an orbit has a certain amount of energy and electrons in different orbits will have different amounts of energy. Since the orbits are quantized, so are the energy values – called energy levels. The electrons fill these levels, from the deepest near the nucleus to the farthest, following certain rules, to be discussed below. An electron could move from one orbit to another by exchanging

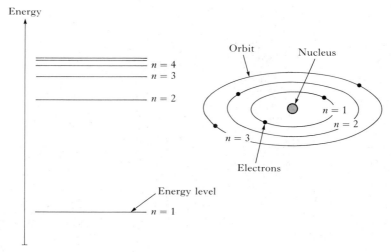

Figure 6.2   Energy levels for electrons in an atom according to Bohr. Notice that the spacing of the energy levels is, in general, not uniform. Discrete amounts of energy are given (or taken) as electrons move down (up) the energy ladder.

quanta, that is, discrete amounts of energy (in this example, by emitting or absorbing light of a certain frequency).

The nicely smooth and jerk-less motion of particles that Newton envisioned was utterly destroyed. The upsetting of the classical, Newtonian world could not have come more swiftly and with more radical consequences. By the end of the twenties, most of the new laws had already been written. This was quite remarkable since some of the old laws of physics had been considered unassailable just a few years before! Some of these changes were so radical that more than a few scientists felt very uneasy about this new physics.

Albert Einstein wrote in 1912:

The more success the quantum theory has, the sillier it looks.

Some of the issues Einstein was alluding to in his quote are at the foundation of quantum mechanics and are still the object of debate today. We cannot follow these discussions in depth here, since they require an advanced background in physics. Rather we will sketch the most important features of quantum mechanics which are necessary for understanding superconductivity.

Quantum mechanics is not just a modification or improvement on previously existing conceptual models; it is radically different from what came before and gives explanations to phenomena that otherwise would be considered either extremely odd or counter-intuitive. For example, experiments indicated that matter some-times exhibited a wave-like behavior rather than the corpuscular (or billiard ball) one we are all familiar with when observing the motion of macroscopic objects, from a tennis ball all the way to a planet. It is indeed hard to think that an object which we can touch can be considered a wave. But, for argument's sake, let us believe for a moment that matter does behave like a wave. If we suppose that the position of an electron or an atom is represented by a wave, then the following question arises: where is the electron (or the atom)? This question can be difficult to answer, since the electron or atom is spread throughout the length of the wave and it is impossible to assign a position to it as we do in 'classical', that is, Newtonian, physics. If we describe an electron as a wave, see Figure 6.3 for some examples, we cannot pinpoint the position of the particle; we can only be reasonably confident of finding the particle in that particular region of space where the wave is

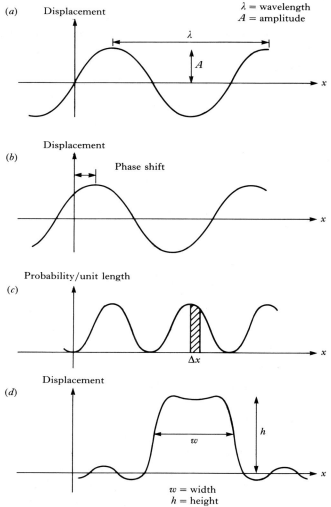

Figure 6.3(a) Snapshot of simple wave of amplitude A and wavelength λ. (b) Same as above, but the displacement is different from zero for $x = 0$. Notice that maxima and minima are shifted with respect to wave in (a). (c) Probability per unit length versus x. This curve is obtained by multiplying curve (a) by itself. Shaded area gives the probability of finding the particle in a neighbourhood around $x = x_0$. (d) Another type of wave, a pulse. In this case the wave depicts a particle whose location is spread over a distance w.

confined. In other words, we are no longer certain where the particle is, but, through mathematical tools that scientists such as Erwin Schrödinger developed in the twenties, we have to calculate with a certain degree of confidence where to find it. A similar argument can be made about the velocity of the particle.

6.2 The new rules of the game

At the core of quantum mechanics is the underlying premise that we cannot specify with precision the position *and* velocity of a particle *at the same time*, as we would normally do, given the appropriate instruments, in classical mechanics. We are not just saying that instruments are imperfect; we are saying much more: that it is *impossible* (irrespective of the quality and type of instruments) to know the position and velocity of the particle at the same time. It is a very unsettling proposition and, no matter what the level of one's physics knowledge, one always feels a bit queasy thinking about the consequences of this new way of looking at the world. Werner Heisenberg translated this uncertainty in the following way (known now as the Heisenberg uncertainty principle):

$$\mathrm{d}X\,\mathrm{d}P > h,$$

where $\mathrm{d}X$ and $\mathrm{d}P$ are the uncertainties in the position and momentum (velocity times mass) of the particle and h is Planck's constant. Thus, the product of those uncertainties cannot be less than a given amount. We can make $\mathrm{d}X$ small, but then $\mathrm{d}P$ has to be large to satisfy the relation above, and vice versa. As stated before, these uncertainties are not measurement errors, but come about *because* we sufficiently perturb the system with our measurements that the position or the velocity (or both) can no longer be known precisely. The relationship above is the new rule of the game.

Many scientists have worried about the philosophical implications of statements like the ones presented above. Here we will use the results of quantum mechanics without concerning ourselves with wider meanings and implications. Although at first quite unsettling, there is something positive about all this: quantum mechanics has worked very well, allowing scientists to explain many phenomena which couldn't have been accounted for by previous 'classical' theories. This is indeed a good enough reason to keep quantum mechanics!

Coming back to our problem, we could assign to the particle a mathematical function (called in jargon the 'wave-function') the square of which gives the probability of finding the particle in a given volume.† As in the case of a wave, the wave-function is characterized by an amplitude and by a phase. The amplitude and phase might depend on the location in space that we are considering. Let us take a simple case, a sinusoidal wave, which has the simple oscillatory pattern shown in Figure 6.3(a). Here the amplitude is constant throughout space and represents the maximum height, while the phase, which is also constant in this example, gives information on where, from a reference position, the maximum will occur (compare Figures 6.3(a) and (b)). In an analogy with waves or ripples in a pond, the height of the ripple above the quiescent water will be the amplitude; the phase is, at a fixed instant in time, where the maxima of the ripples occur with respect to a given reference point, such as the origin of the ripples.

It is easy to see that interference effects, as in physical optics with ordinary light waves, are indeed possible and sometimes truly spectacular. For example, let us consider an electron coming toward a double aperture and ask where the electron will hit a screen placed behind these openings, see Figure 6.4. If an experiment is carried out, what is observed on the screen is a pattern of intensity (the number of electrons falling on a given detector in a given period of time) similar to the one we would have measured if we had replaced the electron with light. In other words, we obtained an interference pattern, where the maxima of intensity represent the places in which it is most probable to find the electron, and the minima of intensity represent points where it is least likely to find it. In order to produce this pattern we have to work with wave-functions, as shown in Figure 6.3(a). In essence quantum mechanics tells us that at the detector (or screen) we have to sum all the wave-functions and *not* the intensities (which are

---

† A mathematical function returns a value after a certain set of operations – such as multiplication, summation etc. – have been performed on an input variable or quantity. For example $y = 5x$ returns a value (y) when we assign a value x. This value y is obtained by performing certain mathematical operations on the variable x, in this case we multiply it by 5; $y = x^2$ (reads: x 'squared') returns the value y when we multiply x by itself.

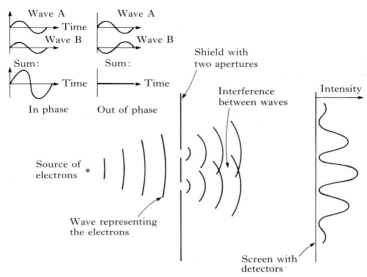

Figure 6.4   Electrons, represented here by a wave, impinge on a shield with two small apertures. Each aperture produces its own wave and the waves from the two apertures interfere. Since each wave has positive as well as negative values for its amplitude, the sum of the two waves on the screen can be positive, zero, or negative. What is plotted on the screen is the recording of the detectors; the detectors measure the intensity of the resulting wave, that is, the square of the sum of the two individual waves. Notice that the intensity is highest in the region between the apertures. (This result couldn't have been obtained using classical mechanics.)

given by the squares of the wave-functions). In fact destructive interference (which is zero intensity) can occur at certain points on the screen because the positive part of the wave-function from one aperture can be cancelled out by the negative part of the wave coming from the other opening. Notice that there is a significant probability of finding the electron in positions which couldn't be reached if the electron behaved as a classical particle. For example, in Figure 6.4 the maximum of intensity occurs in the part of the screen which is blocked by the wall between the apertures! This phenomenon is easily observed using visible light and two small apertures; most lecturers of elementary courses in physics routinely perform this demonstration in class. The novelty here is that the same phenomenon occurs for particles, such as electrons, atoms or neutrons. The success of quantum mechanics lies in the fact that it successfully predicts what is actually observed in an experiment such as the one shown in Figure 6.4.

In quantum mechanics a law, called the Schrödinger equation, has the role of Newton's second equation of motion (Newton's second law). The physicist's job is to solve this equation. There is an important difference, however. The Schrödinger equation doesn't apply to the position of the particle (or derived quantities such as velocity and acceleration), but rather to the wave-function described above. The square of the value of the amplitude (for the mathematically inclined: it is the square of the modulus) gives the probability of finding the electron at that position. In other words: we cannot say that this electron, with a given energy, will be there at that time, once we know the forces acting on it; because of the statistical nature of quantum mechanics, we can only say that the electron will be there with a certain probability which we can calculate using the Schrödinger equation. As in the example above, the wave-function can have positive as well as negative values. However, the probability, which is proportional to the square of the wave-function, is always a positive quantity, see Figure 6.3(c).

One might ask: which is right, quantum or classical mechanics? And if quantum mechanics is right, how do we explain why classical mechanics has worked well in so many instances and for so many years? Quantum mechanics is the correct theory for now, though it might be supplemented by something else later. However, in most instances the odd behavior of microscopic particles, such as electrons or atoms, is not apparent in macroscopic objects composed of billions upon billions of them. As stated earlier, there are a few exceptions in which quantum behavior is clearly seen in macroscopic or ordinary objects, and superconductivity is one of the most important of these.

6.3   Quantum mechanics and superconductivity

What is the relationship between quantum mechanics and superconductivity? Onnes thought that the new mechanics, at that time in its infancy, could hold the key to explaining superconducting phenomena. However, physicists in the twenties and thirties had a very difficult time trying to explain superconductivity using quantum mechanics. Because of their considerable technical complexity we cannot describe these theoretical models in detail; nonetheless, below we will try to illustrate in a simplified way some of the problems scientists had to solve.

In metals, the outer electrons of each atom, which are responsible for electrical conduction, are very delocalized. In simple words, it is probable that these electrons will *not* be found near their own atoms, but, rather, almost anywhere in the solid with roughly equal probability. Felix Bloch showed that these electrons can move freely through the crystal undeflected (another typical quantum phenomenon). However, if there are impurities or if the atoms move around their positions because of thermal agitation, the electrons will collide with ions (recall that ions are atoms missing some electrons) and other electrons and an electrical resistance will be measured. Experiments have shown that, as the temperature goes down, the electrical resistivity decreases smoothly and then levels off at a limiting temperature greater than zero. Describing the electrical resistivity in terms of the collision of electrons is not easy, since in order to solve the problem one has to consider the simultaneous interactions of so many particles. Theories have been developed that use quantum mechanics to explain how electrons behave in matter and how their motion gives rise to thermal and electrical conductivities. But all efforts at that time to explain the perfect conductivity aspect of superconductivity using quantum mechanics failed. Something was missing.

6.4 The giant atom

As was true of most of the pace-setting discoveries in science, several clues led toward the correct explanation. Fritz London was one of the first to make a seminal connection between quantum mechanics and superconductivity. In an important meeting of the Royal Society of London in 1935, he conjectured, correctly, that the diamagnetic properties of superconductors were central to a successful theory of superconductivity. He said also that the superconducting state could be described by thinking of the superconductor as a giant 'atom' with electrons whirling around its periphery and producing the shielding currents responsible for the Meissner effect, that is, a zero magnetic field inside the superconductor. The essential point in his reasoning was that the superconductor behaved as a single object not as a collection of atoms. One could describe this giant atom by imagining that there were links between all its parts, since for macroscopic currents to be produced, a macroscopic order (or correlation) among all the electrons throughout the body had to be maintained. This kind of order, as it turns out, can be described by a wave, or, to be more precise, by a wave-function, as an ordinary atom would be. When one considers that most of the applications of quantum mechanics had involved microscopic objects, London's idea to describe a macroscopic piece of matter with a *macroscopic* wave-function was quite unusual, since there was plenty of evidence to suggest that macroscopic chunks of matter at ordinary temperatures behaved as a collection of *uncorrelated* atoms.

Although this model was not intended to be taken too literally, it was the first to propose that the electrons of the actual microscopic atoms making up the 'giant atom' were behaving in unison, or coherently, in order to produce the supercurrents. Scientists were again on the right track. Another seminal London conjecture was the presence of an energy gap in the allowed energy values of the electrons in a solid. In order to understand this latter hypothesis, which turned out to be essentially correct, we have to examine another peculiarity of the atomic and subatomic world.

6.5   More clues

Each elementary particle, such as an electron or a proton, possesses an intrinsic spin; this is a property which can be described using quantum mechanics, but has no analog in classical mechanics. If we really want to make an analogy with classical mechanics, the spin can be visualized as the rotation of a particle around its axis as in a top (see Figure 6.5). This spin, usually given as a fraction or an integer of a universal constant called Planck's constant (the same one we encountered in the quantization of oscillators), is an intrinsic property of the particle, and doesn't depend on its interaction with other particles. For example, an electron has a spin $\frac{1}{2}$ and a particle with this spin is often called a fermion because it obeys certain statistical rules laid out by the Italian physicist Enrico Fermi. Fermions are particles whose spins are an odd multiple of $\frac{1}{2}$. Other particles, whether elementary or

(a)

Energy

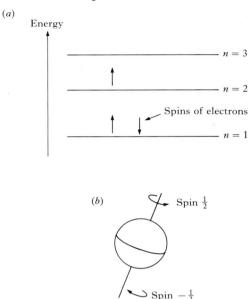

——————————————————— $n = 3$

——————————————————— $n = 2$

Spins of electrons

——————————————————— $n = 1$

(b)   Spin $\frac{1}{2}$

Spin $-\frac{1}{2}$

Figure 6.5(a)   Arrangement of spins in an atom. Not more than two electrons with opposite spins are allowed to occupy an energy level. (b) The classical analog of spin. Only certain discrete orientations of the spin are allowed in quantum mechanics.

composed of even smaller particles, can have other spin values. Rules say that if we fix an axis of arbitrary direction in space, the spin of a given particle on that axis will take only a specific and restricted number of values. In other words, the spin of a particle, referred to an axis in space, is quantized. For example, an electron, which is a 'spin $\frac{1}{2}$' particle as we have seen above, can have only two spin values *referred* to a given axis, $+\frac{1}{2}$ and $-\frac{1}{2}$. If we were to use a classical analogy as before, we could say that a spin $\frac{1}{2}$ particle rotates counterclockwise, a spin $-\frac{1}{2}$ particle clockwise.

The total spin of a compound particle, one composed of several elementary particles (for example, an atom made up of electrons, protons and neutrons), can have only certain 'quantized' (discrete) values. The value of the spin of the compound particle can be computed using specific rules; roughly speaking, these rules say that the spins of different types of particles should be added independently and the grand total computed by adding the contributions of the different particles. Notice that a particle composed of fermions can have a total spin which is an integer. This type of particle is called a boson since it obeys statistical rules proposed by the Indian physicist Satyendra Bose. For example, a helium atom, which has two electrons (with opposite spins), two protons and two neutrons, has a spin of 0 and therefore is a boson. Another type of helium atom, which has one less neutron and is much less abundant in nature, is a fermion, since its total spin is an odd multiple of $\frac{1}{2}$.

The Pauli exclusion principle, named after the physicist Wolfang Pauli, says that two fermions, such as two electrons, cannot occupy the same energy level if they have the same spin. To put two electrons on the same energy level, if one has spin $\frac{1}{2}$, with respect to a fixed axis, the other must have spin $-\frac{1}{2}$. What if we have three electrons, as in the lithium atom? We then put two electrons in the lowest available energy level (we can find this energy level by solving the Schrödinger equation) with opposite spins; the third one is placed in a higher energy level. Remember that energy levels are also quantized, therefore only a discrete number of the energies are available, see Figure 6.2. If we have a solid, specifically a metal, we proceed in the same way (Figure 6.5). The first two electrons go in the first level, the next two (always with opposite spins) in a higher energy level, the next two in an even higher energy level and so on. In reality the rules are a bit more complicated. Because each atom contributes one or more of its outer electrons, we have to arrange a very large number of electrons. The energy level in which the last two electrons are placed is called the Fermi level. At a temperature of 0 K, all the electrons have energies below or at the Fermi level; the energy levels above it are empty. These electrons constitute 'the Fermi sea', so called to convey the image of a fluid of electrons with an upper surface. At room temperature there will be a few electrons which momentarily have enough energy to visit some of the levels above the Fermi level; to keep the analogy with the liquid we can say that these electrons 'evaporate'. In a metal at room temperature many electrons populate these energy levels above the Fermi sea; it is easy to move them around by using an electric field which produces an electrical current. For this reason, metals conduct electricity quite easily, while insulators cannot because they don't

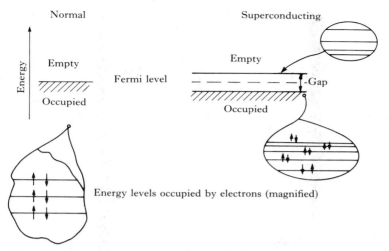

Figure 6.6   Energy level in a normal metal (left) and in a superconductor (right); the spacing of the levels has been exaggerated. Notice the increased density of energy levels near the energy gap. At $T = 0$ K, only the energy levels below the Fermi level are occupied by electrons.

have easily accessible energy levels above the highest occupied energy level.

Coming back to London's suggestion of an energy gap, we can explain it as follows. The energy gap separates the ground (or lowest) state of the system (in which all the electrons occupy the lowest energy levels compatible with the rules mentioned above) from the excited state (in which at least one electron has acquired some extra energy), see Figure 6.6. Obviously the latter has an energy higher than the ground state. For superconductors, experimental evidence and other conjectures pointed to the fact that these two states (ground and excited) were not contiguous, but separated by a small amount of energy. In such cases, electrons were not easily available for conduction, since some energy had to be given to them so that they could jump over this *energy gap* into the empty states.

The next clue came from experiments. Several scientists, starting with Onnes, tried to see which properties changed and which didn't when a material became a superconductor. Was superconductivity a phenomenon caused by the electrons only, or did atoms (or ions) play a role in it? The answer to this question was given by measuring the critical temperature of superconducting materials composed of atoms with the same electronic configurations (that is, the same number of electrons distributed in the same way with respect to the nucleus) but different masses in the nucleus (due to the different number of chargeless neutrons). Atoms which differ by the number of neutrons only are called isotopes of that given element. Typically, one element has a preponderant isotope. We have already mentioned that helium can come with two neutrons, the most abundant type or 'isotope', or with one neutron, a much rarer 'isotope of helium'. It was found, in 1950, that the critical temperatures of different isotopes of a given superconducting material were different, thus proving that the mass of the nuclei had something to do with superconductivity. But in what way?

Before we present what is now accepted as the most successful and complete explanation of superconducting phenomena, a theory formulated by John Bardeen, Leon Cooper and Robert Schrieffer in 1957, we have to consider how atoms move in solids.

As a solid is brought from lower to higher temperatures, its atoms, initially all arranged in regular arrays in a perfect crystal, perform larger and larger oscillations around their equilibrium positions. The interactions between atoms are complex and are ultimately responsible for holding the atoms together in the solid. Physicists like to simplify things so that they can build models to study the phenomena under scrutiny better, and we will follow their example. We can model these complex interactions by imagining that atoms are connected to each other by springs, see Figure 6.7. What happens when an atom is plucked away from its rest position? The atom, not unlike the string of a guitar, will vibrate around its rest position, and in doing so disturb its neighbors. The disturbance will propagate from atom to atom through the springs, like a domino ripple, or a wave. Quantum mechanics tells us that these oscillations cannot have any arbitrary frequency; rather (remember Planck's hypothesis), they are 'quantized', which means that only frequencies which are multiples of a fundamental frequency can propagate. This is true for a musical instrument too; here the length of the string determines which vibration modes can be excited, see Figure 6.7. The numerical value of this frequency depends on the type of material

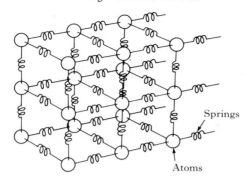

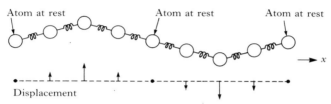

Figure 6.7 A solid can be thought of as being composed of atoms that interact with their neighbours via springs. The coordinated motion of atoms away from their rest position forms a wave (or phonon) as shown. The wave moves toward the right, while the individual atoms move up and down around their rest position.

we are considering. Instead of speaking of waves we will call these vibrations 'phonons' to remind us that the vibrations are quantized. From now on, the words 'phonons' and 'vibrations' will be used interchangeably.

Do phonons interact with electrons? They do in a metal; the higher the temperature, the more vibrations (phonons) are present and the interaction between phonons and electrons produces a higher electrical resistance. Obviously this doesn't happen in a superconductor, since the electrical resistivity is zero until the critical temperature is reached. Why? These were the crucial issues physicists were working on just after the Second World War.

7

Superconductivity explained!

7.1 The mattress effect

THE FIFTIES were years of great synthesis for superconductivity. The phenomenological theories of the thirties had not been able to furnish an explanation of it at a microscopic level (using electrons and atoms), although they did reorganize what scientists knew at that time and did suggest provocative relationships between different facts. It was clear, for example, that perfect diamagnetism was an important manifestation of superconductivity.

John Bardeen gave a candid assessment of how many theorists working in the field felt:

It seems to me that most of those who thought long and hard about superconductivity prior to the discovery of the Meissner effect in 1933 never got over an inner feeling that the really fundamental property of a superconductor is infinite conductivity or persistent currents, and this colored the way they thought about the subject in future years. While an adequate theory must explain both aspects, the diamagnetic approach has been the most fruitful in indicating the nature of the superconducting state.

Ultimately, it was again Fritz London's intuition that showed the way. By the late forties and early fifties, it was clear that to obtain London's wave-function for a macroscopic body a new type of interaction between electrons had to be postulated. We know from elementary physics that like charges, such as two electrons or two protons, repel each other, while unlike charges, such as a proton (positive charge) and an electron (negative charge), attract each other. This is called Coulomb's law and was formulated more than two centuries ago by Joseph Priestley and Charles Coulomb. What seemed necessary to solve our problem was an *attractive force* between electrons. How was it possible to attain this?

John Bardeen started to work on this problem in the late thirties, but it was only after the Second World War that his interest in superconductivity, temporarily diverted by the study of the physics of semiconductors for which he won his first Nobel Prize in Physics, was reawakened (Figure 7.1). He won a second Nobel Prize in Physics (a truly unusual event) for his work on super-conductivity, as we will see shortly.

Bardeen wasn't an ordinary scientist. As a child prodigy, he jumped several grades at school. He didn't quite fit the profile of an 'enfant prodige' either: he had a keen interest in golf, swimming and the social life of a fraternity club. His style of working employed a 'from the bottom up' approach. He paid a great deal of attention to experiments and never tried to bend experimental findings to fit his models, a common disease among many theoreticians of the 'from the top down' school.

Bardeen started to think about superconductivity again just before the Second World War; he believed that the interaction of an electron with the vibrations of the solid (called the electron–phonon interaction) and the energy gap were all essential in an explanation of superconductivity. The catalytic event in the development of a microscopic theory of superconductivity was the discovery, in 1950, of the isotope effect which showed that the critical temperature was related to the mass of the atoms of the solid. This confirmed the importance of the interaction between the electrons and the vibrations of the atoms. Few were the theorists who thought along these lines. After all, the electrons have less than $1/2000$ of the mass of the nuclei, which move very sluggishly, and move so rapidly that it seemed to most researchers the two motions had little in common.

Another theorist, Frohlich, who was also working on this problem just after the War, believed that:

... prior to my introduction of phonon-induced electron interaction it seemed ridiculous to assume attractive forces between electrons, which is probably the reason why the theoretical treatment of superconductivity was held up for so long.

We now come back to the question of how it is possible to have an attraction between two electrons when we know they should repel each other. The core of the explanation is the 'mattress effect'. Consider a heavy ball rolling fast on a soft mattress, see Figure 7.2. The mattress will bend downwards or sink where the ball is. If the ball rolls fast enough, the springs will not have time to relax back to the original position immediately after the ball has

Figure 7.1   John Bardeen (right) with his colleagues Walter Brattain (left) and William Shockley (center). They shared the Nobel Prize in Physics for the invention of the transistor. (Courtesy of AIP Niels Bohr Library.)

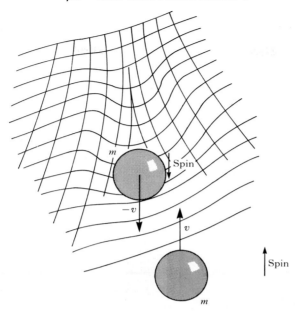

momentum = velocity × mass

Figure 7.2   Balls on a mattress. A ball with mass m and velocity $-v$ has just created a depression in the mattress, the depression attracts the ball with velocity $+v$. According to Cooper, the most favorable condition occurs when the two electrons have opposite momenta (that is, opposite directions but same amount of speed and mass) and opposite spins.

passed, but instead will take some time to do it. Another ball travelling nearby might encounter the depression and fall into it. We can say that the two balls have 'interacted' because they have felt the presence of each other, although with the mattress as an intermediary. Since the second ball is pulled to where the first was, we can say that this interaction is attractive. To come back to our problem, substitute electrons for balls and the solid composed of sluggish ions for the mattress, and we will obtain an attractive interaction between electrons.

Of course saying this is possible is not equivalent to saying this is what really happens, since other forces (such as Coulomb's!) and phenomena compete with this one. Physics is sometimes hard to do because one has to sort out, with calculations, which of several causes is the most preponderant in giving some observed effect. In fact, although all the building blocks were there, it took several more years to figure out how they could be seamlessly assembled. Bardeen worked with Leon Cooper, a physicist familiar with sophisticated mathematical tools, and Robert Schrieffer, a brilliant graduate student of his (see Figure 7.3).

At a meeting of the American Physical Society, Bob Schrieffer recalled what it was like to work for John Bardeen, who had recently passed away. As mentioned at the beginning of this chapter, Bardeen, although a theorist, also had a laboratory in semiconductor physics. He believed strongly that theorists should have an idea of what experimental physics was all about. Taking data is perhaps the smallest, and most pleasant of the chores of an experimental physicist. More often one has to fix parts of

Figure 7.3   John R. Schrieffer (right) and Leon N. Cooper (left) together with John Bardeen shared the Nobel Prize in Physics for their theory of superconductivity. (Courtesy of G. M. Almy, AIP Niels Bohr Library.)

equipment or build and test new instruments. Schrieffer recalled how one day Bardeen went down to his laboratory and discovered that an electrical lead was disconnected from a circuit. Unintimidated, he proceeded to solder it, while all the laboratory assistants, pretending to go on with their own projects, were anxiously peeking at him to see how he was doing (and he did a fine job).

Bardeen thought that every graduate student, including the ones wishing to work in theoretical physics, should work in a laboratory for a year or thereabouts, so he sent Schrieffer into a laboratory for a year. The stint was actually rather brief. One day, while welding some pieces of metal together in a hydrogen atmosphere, Schrieffer caused a small explosion. It was immediately decided that Bob would henceforth work on *pure* theory.

In the meantime, Leon Cooper (Figure 7.3) found that if two electrons have the same velocity, but are moving in opposite directions, with opposite spins, the attractive part due to the substrate-mediated interaction is stronger than the Coulomb repulsion. Thus, a net attractive interaction is obtained; these two electrons form a pair, and this is now called a Cooper pair. The analogy with the balls rolling on the mattress is more accurate than one might imagine at a first reading. In fact, neither the electrons nor the balls physically form a pair adjacent to each other. It would be more appropriate to think of the two electrons as a pair characterized by their momenta (that is, velocity times mass) which are equal in magnitude but opposite in direction. In physics jargon this is called a pair in 'momentum space', since the variables which are relevant are the momenta and not the positions of the electrons. But if we had to visualize this pair in 'real' or geometric space, we would see the two electrons about 2000 angstroms apart (the distance between two atoms in a solid is between 2 and 3 angstroms or 1/50 of a millionth of a centimeter). Therefore, the electrons in the pair are not very close, in fact there are millions of other pairs in between the two electrons of a given pair! It seems quite strange that this can happen. We might view this as being similar to what happens in a discotheque, where two partners dance synchronously several meters apart, while other couples slide in between.

It is time now to recall what we said about electrons in the previous chapter. Because of the Pauli exclusion principle, not all electrons can have the lowest energy. By following the rules, we put electrons two by two, with opposite spins, into increasingly higher energy levels. The last electrons we place are the ones responsible for electrical conduction; their energy level is called the Fermi level. In arranging the electrons on different levels, we have assumed that they were far away from each other, so that we could neglect their repulsive interaction. What Leon Cooper discovered was that electrons can feel an attraction for each other. When this occurs the electrons near the Fermi level get redistributed a bit; his calculations showed that some electrons have their energy pushed down from the Fermi level, while others have their energy pushed up. Since the number of electrons is the same, some of the energy levels, especially near the top, are closer to each other (denser) than the others, see the right panel of Figure 6.6. A gap, or absence of energy levels, results near the Fermi level; the theory of quantum mechanics says that no electrons are allowed to have values of energies which are in the gap. The presence of this gap, as was correctly intuited many years before, is responsible for many of the properties of superconductors, and should be considered a cornerstone of the microscopic explanation of superconducting phenomena.

Those who are more familiar with the properties of matter might suggest that there is a whole class of solids which have an energy band gap and yet don't show superconducting properties. They are the semiconductors† which nowadays form such an important part of technology, being the raw material from which computer chips and advanced electronic devices are made. The reason why the role of the energy gap in a semiconductor is different from that in a superconductor is subtle. Here we can say that a key result from Leon Cooper's calculations was the discovery that the energy gap is due to the presence of tightly correlated 'Cooper' pairs and that each pair has two electrons with equal but opposite spins. No such correlated motion is present in semiconductors. An important consequence of this crucial difference is explained below.

---

† Silicon, germanium, and gallium arsenide are examples of widely used semiconductors. These semiconductors have values of electrical resistivity (10^3 ohm × meter) which are between the metals (10^{-8} ohm × meter) and insulators (10^{10} ohm × meter or even higher).

7.2 The BCS theory

With all these elements of the puzzle coming together, John Bardeen, Leon Cooper, and Robert Schrieffer worked non-stop to calculate key quantities that could be compared with experimental results. As late as the Fall of 1956, the end was not in sight. Schrieffer was worried that the solution to the superconductivity problem would take several more years of work. He quietly started to work on another problem in order to have something ready for his graduation. Instead, in early 1957 came the breakthrough. Some technical difficulties that prevented Bardeen, Schrieffer and Cooper from proceeding with their calculations were overcome. All of a sudden (and this is not unusual in science!) every piece of the puzzle fell into place. John Bardeen recalls that

... we were continually amazed at the excellent agreement obtained [between our model and experimental results]. If there was serious discrepancy, it was usually found on rechecking that an error was found in the calculations.

In the spring of 1957 a paper detailing their theory of super-conductivity was submitted to the journal *Physical Review*.

Let us see how the theory works in explaining two key superconducting properties: the persistence of electrical currents and the specific heat anomaly. If we set up an electrical current in a superconductor (in a ring, for example, as Onnes did many years before) the current will (contrary to what happens in the everyday world) keep on flowing after the battery which provided the initial push for the electrons has been removed. We know that the electrical resistance is due to the scattering of electrons with impurities and lattice vibrations (phonons) which help randomize the motion of the electrons. Superconductivity doesn't suppress the scattering; the theory just says that the pair is scattered as a single object without being torn apart. For electrons to be effectively scattered so as to produce an electrical resistance, the pair has to be broken up; this operation requires an energy at least as large as the energy gap. Each pair carries a net momentum imparted by the electric field (see Figure 4.4); to change that momentum some energy is required. Occasionally, at a temperature below the critical temperature, this amount of energy will be available to a pair to change its momentum. To stop the current, however, *all* the pairs and not just one have to be stopped, which requires a considerable concerted effort (this being the difference with respect to semiconductors). Therefore, this extra occasional energy available for breaking up one or a few pairs, a fluctuation as it is called in jargon, will produce a tiny change in the overall current, but it will not be sufficient to make it decay to zero. When the temperature is raised and approaches the critical temperature, numerous pairs start to break up since the lattice has enough energy (because of heat or, equivalently, phonons) to overcome the energy gap.

At the transition temperature all the Cooper pairs are destroyed. Theoreticians working in superconductivity have come up with special names for this process; they see the breaking of Cooper pairs as the creation of excitations. These excitations, which consist of electrons which were previously bound in a pair and now have become free, are called quasi-particles. As the temperature is raised, more pairs are destroyed and more quasi-particles created. At any given temperature greater than 0 K, we have electrons bound in pairs and quasi-particles. This indeed brings to mind the two-fluid model introduced earlier; the quasi-particles represent what we called the normal component, while the remaining Cooper pairs represent the superconductivity component. As we said before, sometimes approximate, phenomenological models contain the right physics after all!

A similar argument can be used to explain the specific heat data, that is, the fact that these data imply the existence of an energy gap for the electrons. The specific heat, as we have seen, is the amount of heat we have to give to a gram of a substance to increase its temperature by 1 K. The higher the specific heat, the harder it is to raise the temperature in a substance. What contributes to the specific heat? The ions in a crystal (the atoms minus the conduction electrons) are in part responsible for it, but at these low temperatures, and for metals, the conduction electrons are mainly responsible for the thermal properties of the solid. In a superconductor we need to give a minimum amount of energy to each pair to break it up; below that amount of energy no pair is broken. Thus at the transition temperature, when there is a very large number of pairs ready to break up, a considerable amount of

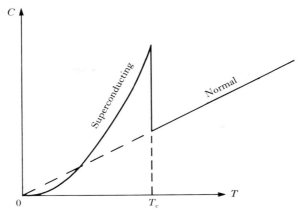

Figure 7.4 Specific heat (C) due to electrons versus temperature (T). The dashed line is the specific heat of a normal metal. T_c is the critical temperature.

energy is required, making the specific heat large, as indeed is observed, see Figure 7.4. Other experimental techniques were used to probe this gap directly. For example, Glover and Tinkham in 1957 just before the publication of the BCS theory used infrared light (that is, light with frequencies below the visible red) on superconducting samples. In these experiments, if the light had an energy (recall energy = frequency times h – Planck's constant – see Section 4.2) greater than the superconducting gap, it would break up the Cooper pairs and be absorbed by the sample. Tinkam and others showed that there is indeed a gap in samples of tin, which is superconducting below 3.7 K.

The basic elements of the theory, now universally known as the BCS (Bardeen, Cooper, Schrieffer) theory, were published in 1957. Despite its rocky start (not everybody accepted this explanation when it appeared) the theory proved to be enormously successful. We can speculate that the reason for this success was that it didn't just explain well recognized experimental facts but spurred experimental and theoretical work. For their contribution, Bardeen, Cooper, and Schrieffer were awarded the Nobel Prize in Physics in 1972.

Rapid progress in experimental and theoretical research on superconducting properties and applications followed. Before we look at some of the applications of superconductivity, we have to consider two important scientific discoveries, one by Ivar Giaever and the other by Brian Josephson.

7.3   Piercing barriers

We spent some time in an earlier chapter describing the meticulous preparation with which Kamerlingh Onnes carried out his experiments. A different style was used by Ivar Giaever, who did his experiments, for which he was awarded the Nobel Prize in Physics, while he was literally still taking courses in physics!

Ivar Giaever understood that:

... the road to a scientific discovery is seldom direct, and it does not necessarily require great expertise. In fact, I am convinced that often a newcomer to a field has a great advantage because he is ignorant and doesn't know all the complicated reasons why a particular experiment should not be attempted.

Before we examine the contributions of Giaever and Josephson we have to tell of another very unusual and, perhaps, counter-intuitive phenomenon encountered in the atomic world. It is well known in classical mechanics that a particle moving toward an obstacle, such as a wall, will bounce back after hitting it. If we examine a bit more carefully what happens during the collision, we can describe it by saying that the energy of the particle is not enough to penetrate or pierce the wall. What stops the ball going through is the interaction, strongly repulsive, between the atoms of the ball and those of the wall. Physicists like to say that there is an energy barrier that has to be overcome in order to go through the wall. Of course, if the particle has enough energy, because it is very fast and massive as is true of a cannon ball, then it will go through. In this case we say that it has more energy than the energy barrier of the wall. Because energy is conserved, as scientists firmly believe, a particle needs an energy greater than that of the barrier in order to overcome it, otherwise it will be bounced back.†

In the atomic world, things behave a bit differently. Let us take an electron, or another light-in-weight particle, and send it against an energy barrier. This energy barrier could be created, as in the example above, by other electrons in an atom or group of atoms.

† We assume, for argument's sake, that the collision is elastic; that is, no energy is dissipated into a disordered motion of the atoms making up the wall (heat).

(a)

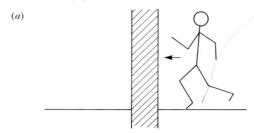

(b)

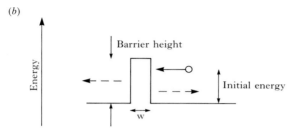

Figure 7.5(a)   A person trying to 'tunnel' through a wall has a very small probability of succeeding. (b) An electron with an initial energy smaller than the energy barrier height would be reflected back if it behaved as a classical particle. If the barrier is thin enough, however, quantum mechanics predicts that the electron has a good probability of going through.

Quantum mechanics tells us that the particle has a non-negligible probability of going through the barrier *even if* the energy of the particle is smaller than the barrier's (see Figure 7.5). This process is called *tunneling* to suggest that the particle tunnels from one place to another not by overcoming the obstacles (energy in this case), but by actually going through them. This sounds like a trick in a science fiction story, but it is true! A person also has the capability of tunneling through walls from one room to another; a calculation using quantum mechanics will tell us that the probability of this event is astonishingly small, as anyone can easily verify in practice. For particles with small masses, and for thin energy barriers (or walls in our example), however, this probability can be important and this is what Giaever exploited.

Giaever was assigned as his first job at the General Electric Laboratories in Schenectady the measurement of electrical conduction in thin films. A thin film has nothing to do with movies, as

Giaever, a mechanical engineer by training, thought when he was assigned this project; rather a thin film is a stack of atomic layers deposited on a substrate. At the same time that Giaever was carrying on his experiments and waiting '... to switch into theory as soon as I acquired enough knowledge', he was taking courses in physics at the Rensselaer Polytechnic Institute in nearby Troy. When he took a course in superconductivity, he soon realized he could use the tunneling of electrons to measure the energy band gap in a superconductor, a notoriously difficult quantity to measure. He took a metal (aluminum) and put on top of it a very thin electrically insulating film and then put a super-conductor (lead) on top of this film to form a sandwich (see Figure 7.6). Because of the insulating layer, classical mechanics would predict that electrons couldn't travel from one metal to the other. But according to quantum mechanics they can (with a certain

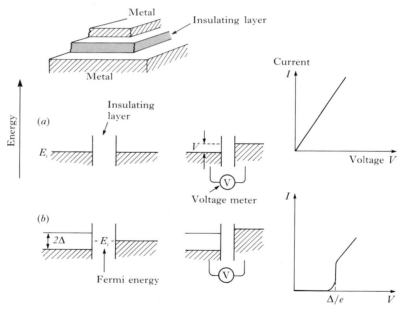

Figure 7.6   Tunneling junctions and *I*(current)–*V*(voltage) curves. A tunneling junction is formed by two metals with a thin insulating layer in between. (a) Two normal metals at the same electrical potential (left) and when a potential is applied across the junction. This device obeys Ohm's law. (b) Same as (a) but with the metal on the left in the superconducting state. Notice the energy gap is 2Δ. The *I–V* curve can be used to measure Δ; 'e' is the charge of the electron.

probability) and this is what he observed. Tunneling was not a new concept in science but it had been used primarily in nuclear physics. Giaever's contribution was to use tunneling to show that there is an energy gap in a superconductor and to demonstrate a way to measure it.

From Giaever's notebook:

Friday, April 22 (1960), I performed the following experiment aimed at measuring the forbidden gap in a superconductor.

He used the following method. If a thin insulator is placed between two metals a very feeble current will go from one side of the sandwich to the other; but if one of the metals happens to be in a superconducting state then, for small applied voltages, there will be no tunneling, and therefore current, because the electron in the metal will not have any place to go, since there will be a gap at the Fermi level in the superconductor where no electrons can stay, see Figure 7.6(a). If the applied voltage is increased, the electron will be able to go into the states of the superconductor which are above the gap, see Figure 7.6(b). Why do we need an insulating layer in between the metals? It is well understood that if we put two metals together they will reach the same electrical potential. Now that we know about energy levels of electrons, we say that the two metals, if in contact, will match their Fermi levels (Fermi energies). If the two metals are at the same electrical potential no current will flow. But if we have an insulating layer in between we can maintain the two sides of the sandwich at two different electrical potentials or Fermi levels. If we do as electrical engineers do to find out the electrical characteristics of a device, we can plot the value of the current and voltage as in Figure 7.6. From this graph (called an 'I–V curve') one can calculate the energy of the gap in the superconductor.

The trick that allowed Giaever to carry out his experiments was that he was able to make an insulating layer thin enough to obtain a measurable tunneling current, and thick enough to prevent metals from coming into contact in any part of the 'sandwich'. Ivar Giaever was awarded the Nobel Prize in Physics with Leo Esaki and Brian Josephson in 1973.

If we have superconductors on both sides of the insulating layer, why can't Cooper pairs tunnel through? They can, and actually

Giaever did this experiment without fully realizing the consequences. In the early sixties, Brian Josephson, then a graduate
student in Cambridge, England, considered what happens to the
Cooper pairs if a very thin layer separates two superconductors.
We have already mentioned that we can consider a superconductor
as a giant atom to which we can assign a wave-function. If we have
two superconductors separated by a weak link, the insulating layer,
then we have two wave-functions, one on the left and the other on
the right, see Figure 7.7. In general, they will have a phase
difference, that is to say, the crests of the wave-functions don't
always match; the crest of one occurs, for example, when the other
is low. When two superconductors are coupled, say because of
their proximity, and there is a steady electric current flowing
provided by an external battery, a current due to Cooper pairs
(supercurrent) is observed to flow across the junction. There is no
voltage drop across the junction. The current–voltage curve is
shown in Figure 7.7(b). Depending on the junction being considered, no voltage is observed across the junction only for values
of the current below a certain value. Above that value of the
current, a voltage suddenly appears; this voltage is related to the
breaking of Cooper pairs and is proportional to the energy gap.
Notice the similarity between the two current–voltage curves for
the Giaever (normal–insulating–superconductor) and Josephson
(superconducting–insulating–superconducting) junctions, except
for the supercurrent at zero voltage.

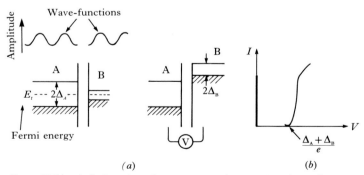

(a)   (b)

Figure 7.7(a)   Left: Junction of two superconductors. A and B, with energy
gaps $2\Delta_A$ and $2\Delta_B$ respectively. Right: when an electric potential is applied,
the energy levels shift. (b) I (current) vs. V (voltage) curve for the junction.

Josephson junctions find a large number of important applications in magnetometry and electronics. These will be illustrated in the next chapter.

In closing, we report what A. B. Pippard, a physicist who made significant contributions in superconductivity in the fifties, said of the BCS theory and the understanding of superconductivity in the mid-sixties. He remarked on the

... overwhelming success of the theory of Bardeen, Cooper and Schrieffer, not only in explaining so very well almost everything that was known about superconductivity at the time the theory was invented, but in providing a framework within which it has been possible to explain the exciting developments that have been made since the theory was first promulgated.

In the same speech given at the close of the 1964 International Conference on the Science of Superconductivity at Colgate University (Hamilton, NY), he said:

This success [of the BCS theory] is so remarkable that I almost believe you would forgive me if I were to say there now remain no problems in superconductivity.

Obviously nobody could have predicted what was to happen a quarter of a century later. Before we consider the latest events we should look at the technological developments in the sixties and seventies which permitted the use of superconductivity in commercial devices as well as in research instruments.

8

Superconductivity-based technology

8.1 Science and technology

THE FANTASTIC PROPERTIES OF SUPERCONDUCTORS that we
have just examined seemed a dream come true for engineers. The
resistance-less flow of electric current could be exploited for
transporting large currents in power lines or in producing very
strong magnetic fields in electromagnets. And, indeed, there are a
considerable number of commercial and scientific applications,
from small devices to large instruments, where superconductivity-
based technology is routinely employed. As we write, the greatest
application of superconductor technology is under way with the
construction of the largest high energy particle accelerator ever
built, the Superconducting SuperCollider (SSC) in Texas. Subatomic charged particles will be hurled around and around the
54 mile circumference of the supercollider. Superconducting
magnets will be used to keep the particles in a circular trajectory.

These new developments notwithstanding, in the past it proved
much more difficult to use superconductors for practical applications than previously thought. As Onnes and others quickly
discovered, there are several obstacles to achieving that goal. The
first is technical and related to temperature. Nowadays, closed
cycle helium refrigerators, working on a principle similar to the

one employed in familiar kitchen refrigerators, are routinely used in scientific laboratories as well as on factory floors to attain a temperature range of 4–10 K, a temperature at which many materials typically become superconductors. However, the use of refrigeration at such low temperature is not cheap nor practical for large-scale applications such as power lines and electromagnets. The second obstacle is more related to the physics of super-conductors. For the superconducting materials we have encountered so far, the passage of even modest electric currents produces magnetic fields large enough to destroy superconductivity. It is common usage to call superconducting materials in which the magnetic field penetrates abruptly into the sample above a threshold or critical magnetic field value (the Meissner effect explained earlier) Type I superconductors. As mentioned before the reason why the magnetic field penetrates the sample is relatively simple. It costs energy to expel the magnetic field from the interior of a superconductor; for example, supercurrents have to be set up on the skin of the material in order to produce a magnetic field which exactly cancels the one from the outside. For low values of the magnetic field, this cost is counterbalanced by the gain in energy that the sample obtains by staying in the superconducting state. However, when the external magnetic field becomes high, the cost of setting up these currents is too large and super-conductivity is destroyed.

This was typical of the superconductors discovered in the early years following Onnes' first experiment in 1911. In general these superconductors are pure elements, such as mercury, lead, and tin. Unfortunately, for a wide variety of applications, from inter-connections in computer chips to wires for large electromagnets, a large current density, or current per unit area of the cross section of the wire, is required. Current densities of the order of 10^6 amperes per square centimeter are often necessary. To give an idea of the value of these current densities, we calculate that in a wire of 3 millimeters diameter this current density would correspond to a current of the order of 10^5 amperes, which is an enormously large current; indeed, a lamp cord, which is approximately the same size, usually carries only a few amperes. Unless special, and often costly precautions are taken to remove the heat generated in the wires, the wires themselves will vaporize.

Since the first years of superconductivity research, the attention of experimentalists has turned to the exploration of other, more complex types of solids which might show superconductivity. Starting with Mendelssohn in England, physicists tried to combine two or more elements (usually metals) to form alloys which displayed unusual properties, not easily attributable to either of the elements. Some of these alloys, proved to be superconducting. More interesting was the fact that they were capable of carrying large current densities; these superconductors are called Type II

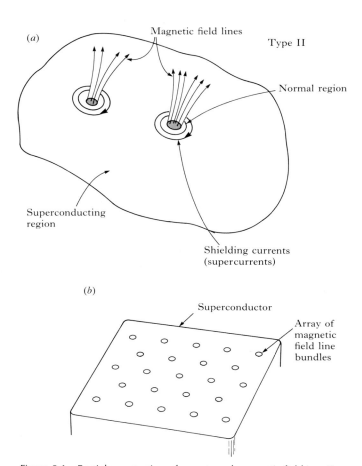

Figure 8.1   Partial penetration of an external magnetic field in a Type II superconductor. The magnetic field has a value between H_{c1} and H_{c2}.

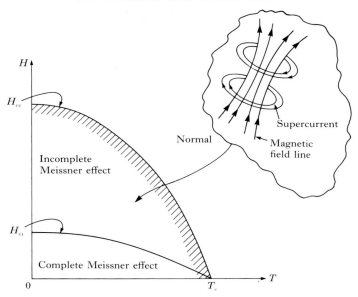

Figure 8.2   Critical magnetic field curves for Type II superconductors. Partial penetration of magnetic flux occurs between curve H_{c2} and curve H_{c1}.

superconductors and they exhibit features quite distinct from Type I superconductors.

For Type II superconductors the considerable energy cost of expelling large magnetic fields, as much as hundreds of thousands of gauss, is reduced by letting some of the magnetic field penetrate the sample. This penetration is not uniform throughout the sample, as occurs when a superconductor becomes normal; instead, in superconducting Type II materials, it proceeds in 'bundles' of magnetic field lines, see Figure 8.1(a). Shielding currents circulate around the bundles of these lines (called vortices). One can think of these bundles of magnetic flux† as bundles of spaghetti; in this analogy the spaghetti would represent the magnetic field lines, that is, the direction and strength of the magnetic force. Inside these bundles the material is normal (that is, not superconducting), while in the other parts, between the bundles of filaments, the

---

† The magnetic flux is the product of the magnetic field strength times the area being considered. In this case the magnetic field is present in the bundles but not outside it.

magnetic field is successfully expelled; these parts are super-conducting.

A Type II superconductor, aside from its different behavior in a magnetic field (Figure 8.2), is essentially similar to a Type I superconductor in which the magnetic field is expelled completely. There is a convenient way to distinguish one type from the other. Two key quantities describe a superconductor: the coherence length ξ (called *xi*), the distance over which the superconductor can be represented by a wave-function, and λ (called *lambda*), the London penetration depth, that measures the distance over which the magnetic field partially penetrates the sample before the supercurrents are able to screen it out completely (Figure 8.3). The supercurrents responsible for shielding out an external magnetic field are localized in this region near the surface of the super-

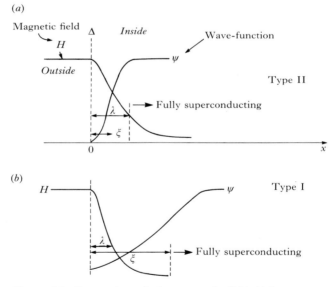

Figure 8.3   Penetration of the magnetic field H in a super-conductor; when the magnetic field decays slowly compared to the coherence length ξ of the wave-function, the superconductor is of Type II. (a) λ and ξ are characteristic lengths associated with the decay of H and the rise of ψ, respectively. The dashed line is the surface of the superconductor. (b) The penetration of H is short compared to the coherence length of the wave-function – Type I superconductor.

conductor. If λ is greater than ξ thermodynamics shows that it is advantageous for the magnetic field to penetrate the sample; the superconductor will then be of Type I. Since these lengths (λ and ξ) depend in different ways on the mean free path of the electrons, Type I superconductors can be made into Type II by reducing the mean free path, for example, by mixing a superconducting metal with another metal. Most metals are Type I and most alloys are Type II superconductors.

The laws of electromagnetism tell us that an electrical current flowing in a superconductor makes the magnetic field lines move. If these bundles were able to move freely, some energy would be spent in moving them, thus effectively making the material electrically resistive. Fortunately, in many materials the bundles are 'pinned', that is, they cannot move because of the presence of defects (imperfections) or impurities in the sample. In such cases little or no energy is lost and the current can flow through the wire effortlessly. How do we know if these 'bundles' really exist? The bundles can be imaged at the surface of the sample by sprinkling it with magnetizable material, such as small iron filings (or dust). These bundles – or vortices – organize themselves in a regular array (or lattice, see Figure 8.1(b)), called the Abrikosov lattice after the name of the Russian physicist who first worked on a theory of Type II superconductors.

8.2 Large scale applications: Superconducting cables, magnets and trains

The most obvious application of superconducting technology is the exploitation of zero electrical resistivity; for example, by making current-carrying wires superconducting, losses due to the resistance of wires which carry electrical power over hundreds and hundreds of miles would be eliminated.† These losses are generally about 5 % of the power delivered. Such a 5 % loss doesn't seem much in relative terms; however, since large amounts of electrical power have to be delivered, that 5 % loss translates into a large sum

† If alternating currents (AC) are employed, such as the ones used to deliver to home outlets, then there will be some small losses even when superconducting lines are used. We recall that in AC lines the voltage (and current) cycle periodically between a maximum and a minimum value.

of money. Why hasn't this technology already been implemented? What are the technical and economic obstacles to building long superconducting power lines?

Let us consider some of the technical, manufacturing and economic problems related to the production and use of superconducting wires. With minor modifications, this discussion can be extended to other large scale applications of superconducting materials.

It is obviously desirable to have materials with the highest transition or critical temperatures. A compound made of germanium and niobium held the record critical temperature of 23 K for almost fifteen years before the advent of high temperature superconductors in the late eighties. Finding a material with a high critical temperature is not easy; BCS theory, which indeed explains superconducting phenomena very well, doesn't give prescriptions for obtaining more desirable materials. Bernd Matthias, who worked with a large number of superconductors after the Second World War and was known for his uncanny ability to discover new superconducting materials, used to make fun of theorists working in superconductivity because they were unable to predict whether a given alloy would be superconducting or not and what transition temperature it would have. Systematic research and intuition allowed him to propose empirical rules for finding new superconductors. Even when guided by these, it is an impossible task to carry out a systematic exploration of the compounds that can be made by mixing together in different proportions the ninety-two elements present in nature. At that time, progress was so slow that many considered the 20 K range for the critical temperature the natural limit for superconductors. This belief was sustained by back-of-the-envelope calculations which indicated that, within the BCS theory, it would be improbable to find transition temperatures greater than 30 or 40 K.

Most wires for demanding superconducting applications are nowadays made of the metals titanium and niobium. Special procedures had to be invented to make the wire flexible and resilient enough to be used in applications which otherwise would have used copper wires. In large scale applications, a superconducting wire has to be overdimensioned to avoid quenches; a quench occurs when a part of the wire for whatever reason (a

Figure 8.4 Microphotograph of the cross section of a composite wire consisting of 2100 niobium–titanium filaments embedded in pure copper. (Courtesy of Supercon, Inc.)

mechanical instability, a momentary loss of a superconductivity in a tract of a wire, a 'hot spot', etc.) returns to its 'normal' or ordinary state of conductivity. In this case the electrical resistivity increases very rapidly (these materials being much more resistive than copper) and a large amount of power is dissipated in a small region of the wire due to Joule heating. If the wire is not designed to handle the surge of power, the wire can vaporize rapidly, thus destroying itself. Furthermore, wires can move, break or otherwise change their characteristics in response to magnetic forces or due to thermal cycling between room and liquid helium temperatures. Such movements can also cause quenches. Figure 8.4 shows the cross section of a superconducting wire. The superconducting filaments are embedded in a matrix of normal metal to avoid quenches and make the assembly more flexible. At Brookhaven National Laboratory a facility has been built to test different types of superconducting wires and ribbons in a 'real life' power plant of 1000 megawatt capacity (1000 times one million watts – enough to power the households of a small city). Figure 8.5 shows the design of such a cable, which is composed of many layers of wires. It is clear that complex design and manufacturing procedures have to be adopted which makes the production of these cables expensive.

When superconducting wires are used in magnets or other applications, they have to be maintained at a temperature lower

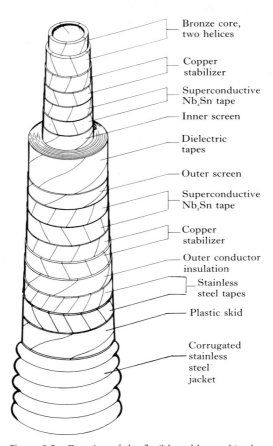

Bronze core, two helices

Copper stabilizer

Superconductive Nb_3Sn tape

Inner screen

Dielectric tapes

Outer screen

Superconductive Nb_3Sn tape

Copper stabilizer

Outer conductor insulation

Stainless steel tapes

Plastic skid

Corrugated stainless steel jacket

Figure 8.5   Drawing of the flexible cable used in the Brookhaven National Laboratory facility. There are inner and outer superconductor tapes. The dielectric tapes are for electrical insulation. Helium is circulated in the stainless steel jacket. (Courtesy of E. B. Forsyth, Brookhaven National Laboratory; copyright: American Association for the Advancement of Science.)

than the critical temperature. Indeed, we already know that at the critical temperature small electric currents or magnetic fields can destroy the superconducting state; as the temperature is lowered higher electric current can be passed in the wires and stronger magnetic fields can be expelled. These and other engineering criteria suggest that the operating temperature should not be much higher than *half* of the critical temperature. Wire made of materials with critical temperatures around 20 K cannot be refrigerated with liquid hydrogen (which boils at 20 K) but instead must be kept around 4 K by using the more expensive liquid helium. Even with the best materials (excluding the new high temperature materials examined in the next chapter), operating temperatures in the 10 K range require the use of liquid helium as a refrigerant. Although technically possible, refrigeration of hundreds of miles of cable at liquid helium temperature poses significant financial and logistic problems.

Paradoxically, the use of refrigerant is not one of the largest expenses in the operation of a long superconducting power line. In reality, other costs probably prevent power utilities from adopting superconducting cables. Wires and ribbons are expensive to manufacture and assemble. Present power lines use overhead aluminum or copper cables which don't require artificial cooling and are relatively easy to maintain. A superconducting power line would need to be buried underground. This is estimated to triple the cost of laying a power line. Furthermore, it would have to be serviced with a coolant, such as liquid helium and vacuum lines for thermal insulation.

The inertia that exists in adopting and implementing a new technology, such as superconducting power lines, on a large scale is due to the difficulty of raising large sums of money for replacing existing facilities, excavation costs and capital equipment. Generally speaking, even when the adoption of a new technology is projected to lower operating costs, the new technology will not be adopted unless these costs are *considerably* less than the present ones. In large scale projects such as sending electric power over thousands of miles, superconductor-based technology might constitute just a small part of a much bigger technological endeavor. Any cost benefits deriving from the use of the new materials might have little weight in determining the budget of the overall project.

Furthermore, large scale applications of any given technology require that the new materials, devices or methods be very reliable, since repeated failures can wipe out the gains that permitted the introduction of the innovations in the first place. As a comparison, we mention coal-fired power plants where old and polluting technologies are still being used despite the availability of cleaner and more efficient ones because of the considerable costs involved in converting such large installations.

These problems notwithstanding, the use of superconducting lines might eventually become advantageous in large and heavily populated metropolitan areas (such as those of New York City or Los Angeles) where bringing in additional aerial high voltage power lines is not feasible or desirable because of environmental or other concerns. The use of cheap and reliable high temperature superconducting cables might tip the balance in favour of employing this technology for long distance transport of electric power, as discussed in Chapter 10.

One of the most remarkably successful applications of superconductor materials has been the production of very powerful magnets. In this application, a wire is wound around a cylinder to form a coil, see Figure 3.3. When an electrical current passes through the wire it produces a magnetic field along the axis of the coil. Very large and strong magnets can be made by using copper wires wound around iron cores, where the iron bar acts as a magnifier of the original magnetic field. On the other hand, these magnets require bulky refrigeration systems because of the heat dissipated by the small but not negligible resistivity of copper. Because of refrigeration problems and because the 'cores' of these magnets saturate, once a threshold magnetic field is reached, it is hard to achieve magnetic fields of more than a few tesla (1 tesla is 10000 gauss or 20000 times the Earth's magnetic field).

In the early sixties several alloys, most notably niobium–titanium and niobium–tin, were discovered to be superconducting with large current density carrying capacity and, perhaps more importantly, to be ductile enough to be shaped into various forms, such as wires or ribbons. Considerable efforts were devoted to making superconducting magnets since with this technology high magnetic fields could be achieved in a small space. We can mention here only a few of the many applications of superconducting magnets in instruments widely used today.

Superconducting magnets are employed routinely in many hospitals in Magnetic Resonance Imaging (MRI) applications. In MRI, a magnetic field aligns the spins of the nuclei in the direction of the magnetic field itself (see Chapter 5 for a definition of the spin of a particle). Then an electromagnetic pulse is given to the nuclei's spins effectively moving the spins away from the original position (see Figure 8.6). When the spins go back into alignment, because the first pulse has ceased, they do so by emitting an electromagnetic signal, essentially a characteristic radio wave, which can be picked up by a receiver. Hydrogen atoms give a good signal; therefore MRI can be used to probe the human body, since tissues of different consistency and water content give signals with different strengths or intensities. In MRI, the role of the superconducting magnet is to produce a very precise and strong magnetic field which helps align the spins of the hydrogen atoms of the tissues. The signal produced by spins as they go back to their original position following the initial pulse is picked up by detectors and analyzed by a computer. Application of the excitation pulse at different angles with respect to the axis of the body to be imaged yields, after extensive computations, a picture of two-dimensional cross sections (or slices) of the body. A similar process occurs in the familiar X-ray imaging (CATSCAN), with the difference that X-rays are most sensitive to bones and other dense materials, while MRI is most sensitive to the soft, aqueous tissues of the body. MRI

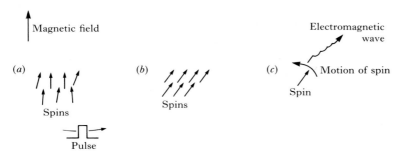

Figure 8.6(a)   Spins are aligned in the direction of the external magnetic field. Because they are not at 0 K, thermal energy randomizes to a certain extent the direction of the spins. (b) Following an electromagnetic pulse, the spins tilt, on average, toward one side. (c) The spins go back toward their original orientation; in doing so, they emit electromagnetic waves which can be detected.

Figure 8.7(a)   Photo of one of more than 1000 superconducting magnets in use at Fermilab's Tevatron facility. These magnets use niobium–titanium superconducting wire and produce magnetic fields of about 4.5 tesla. (Courtesy of Fermilab.)

Figure 8.7(b) The tunnel that houses the particle accelerator at Fermilab. (Courtesy of Fermilab.)

is widely used in hospitals to diagnose tumors, especially in sensitive parts of the body, like the brain, where other techniques cannot be used.

Besides medical research and diagnosis, one of the most interesting and fruitful applications of superconducting magnets is in high energy physics research. There are many other areas of science, such as low temperature physics and plasma/fusion physics, where powerful superconducting magnets are indispensible for research. However, because of the scale of high energy particle colliders, the design, construction and testing of the superconducting magnets employed in these machines constitute a large endeavor; progress and failures have been given ample coverage in scientific magazines and, on some occasions, in newspapers. At Fermilab, near Chicago (see Figure 8.7(a) and (b)), superconducting magnets are used to deflect protons (positively charged particles which are part of the nucleus of an atom) so as to keep them in a circle 4 miles long. The higher the velocity of the particles, the greater the magnetic field required to bend the trajectories of the particles. High energy physicists want to utilize the highest achievable velocities; the higher the velocity of the particles, the higher the energy available when the particles are made to collide with a target. If the energy given to the target is large, there will be a good chance of producing a shower of smaller particles streaming away from the target. The study of these showers allows physicists to study the most basic and ultimate constituents of matter. The higher the energy available, the better the chances of piercing into the ultimate blocks of matter. For this reason, high energy physicists want to build the Superconducting SuperCollider (SSC) which, with its 56 mile long circle, will be able to achieve much higher energies than Fermilab can deliver. With its eight billion dollar plus price tag, it will be the largest and most expensive scientific apparatus ever built. Construction of the tunnel has just begun and prototype superconducting magnets are being built and tested. In the SSC, superconducting magnets will play the important role of keeping the protons on a circular path. Magnetic fields of over 6 tesla (60000 gauss) will have to be produced. Thousands of magnets with stringent performance and reliability specifications will have to be built and assembled in place: a truly amazing engineering project!

Large magnets have found a use in the prototype plants for 'Superconducting Magnetic Energy Storage' (SMES). The facility, originally built for the 'Star Wars' defense program, works as a reservoir of energy. When the electric power grid is under-utilized the SMES plant can absorb and store energy. This electromagnetic energy, stored in large superconducting magnets can be dumped back into the power grid to satisfy increased demand during peak hours. Because of the large construction costs, at present, there is little interest in employing the giant 'batteries' for civilian use.

Finally, we consider the use of superconducting magnets for *mag*netic *levi*tation (maglev, for short). This is, perhaps, one instance in which large scale applications of superconductivity seem closer to becoming reality. Again, in such a large project, the costs of the superconductivity-based technology (essentially the train cars, refrigeration units etc.) represent a small fraction (perhaps 10%) of the costs of laying the railway bed and other infrastructures. In very congested areas, this might indeed be the solution for the problem of rapidly transporting a large number of people over moderate distances (300–500 miles). As a matter of fact, superconducting magnets have been used in test trains in Japan for almost two decades, see Figure 8.8. There are now plans to build the Linear Express, a superconducting train shuttling passengers at 500 kilometers per hour (300 miles per hour) between Tokyo and Osaka.

Let us look briefly at this technology since magnetic levitation trains cannot be considered the evolutionary product of a well established technology; on the contrary they are the fruit of a new emerging technology. Incidentally, non-superconducting magnets could be used in trains as well, as in the German maglev train concept. However, the Japanese prototypes and most of the American feasibility studies employ superconducting technology. Here is how magnetic levitation works.

There are two major lines of design currently pursued by Japanese manufacturers (who, incidentally, are at the forefront of this technology). Both schemes essentially use magnetic forces to lift a train off the railroad tracks. In the attractive levitation train scheme, see Figure 8.9(*a*), a T-shape rail lies in the middle of the two conventional rails. The train hugs the T rail from beneath, where superconducting magnets are placed (the magnets are located on the train). These superconducting magnets generate a magnetic field which attracts the rail; since the rail is rigidly bolted to the ground, it is the train that moves toward the rail, thus lifting itself away from the ground. A delicate balance between downward gravitational and upward magnetic forces has to be achieved by carefully adjusting the magnetic field; otherwise the train would slam into the rail. In the second scheme (Figure 8.9(b)), the repulsive levitation train, there are non-superconducting coils on each side of the railroad bed. Superconducting coils lie just above

Figure 8.8   Superconducting magnetic levitated train in Japan. (Courtesy of: Japan Railways Group.)

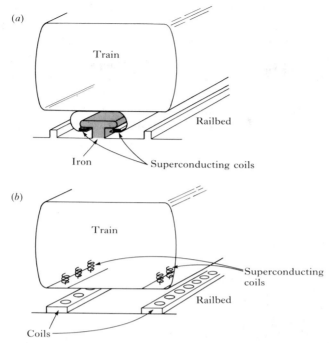

Figure 8.9   Two schemes for magnetic levitated trains. (a) On-board superconducting magnets interact with an iron T-rail. (b) Superconducting magnets on the train interact with normal metal coils on the railbed.

on the train. As the train moves, the magnetic field generated by the superconducting coils on the train passes over the coils on the ground. As we have seen before, according to Faraday's law, whenever the magnetic flux (which is the magnetic field times the area of the coil) is changed, a current appears in the coil (the so-called 'induced current'). This induced current generates a magnetic field of its own which lasts as long as the current flows. Since the coils on the railbed are not made of superconducting material, the current dies out quickly. If the train goes fast enough, it needs to experience the magnetic field of a given coil for a very short period of time. Other methods of propulsion have to be used when starting and stopping. This magnetic field generated by the coils in the railbed has the polarity needed to oppose the magnetic field generated on the train, thus lifting the train from the tracks, as the pole of a magnet would repel the end of another magnet with the same polarity. Incidentally, the same principle is at work in the levitation of a magnet in the celebrated demonstration of high temperature superconductivity, see Figure 8.10.

Magnetic levitation can suspend trains off the railroad tracks. If there is no contact between the train and the ground, how does the train move forward? The reason for using levitation schemes is to reduce friction; in the case of rail transportation, greater stability and speed can be achieved in this way. On the other hand, if there is no friction, no propulsion scheme using wheels or contact with the ground can be used. Here electromagnetic technology is implemented in the linear synchronous motor. Wires in the railroad bed produce magnetic fields that interact with magnets on board the train. If the magnetic field on the track is carefully timed, the combination of attractive and repulsive interactions of the magnetic fields produced in the train and in the tracks propels the train forward as if it were pulled in front and pushed from behind

Figure 8.10   A magnet (small cube) is levitated above a superconducting disc. Electric currents (supercurrents) in the superconducting disc shield the interior of the superconductor from the magnetic field of the small cube and at the same time generate a magnetic field which lifts the magnet above the disc. When the temperature of the disc is raised above T_c the magnet falls onto the disc. (Demonstration Kit by Colorado Superconductors Inc.; photo by G. Vidali.)

by a magnet (push-pull propulsion). A similar principle guides the rotation of an electric motor.

Maglev trains using superconducting technology have been built and tested. Will we be able to jump on them soon? The answer is yes if we live in Japan and believe in the projections of the Ministry of Transportation. Perhaps in five or six years, the first commercial lines should become operative. In the USA, the government has started to award contracts to companies to study various aspects of the maglev transportation concept. According to a study of the National Research Council just released, one of the biggest hurdles in implementing a maglev transportation system in North America is economic. There are doubts whether a maglev system, or even a 'conventional' high speed rail system such as France's Train of Grande Vitesse (TGV), could be ever built in the US and paid for by passenger revenues alone.

There are other uses of superconducting technology, such as for electric power transformers or ship propulsion motors. Essentially the same considerations which were made when we discussed the use of superconductivity in other large scale projects apply here as well.

8.3   Superconducting electronics

On paper it seems that the introduction of superconductivity-based devices should be a boon for high technology electronics. There are two major areas in which superconducting properties can be exploited: in chip interconnections and in electronic gates. In the first case, the rather mundane problem of how to pack so many circuits into so little space translates into the problem of dissipating the heat generated by the electronic circuits before the device burns out. Because of the small dimensions, the heat capacities of these devices are small; in other words, it takes little heat to raise the temperature to a level where the various elements of the circuit (data pathways, transistors etc.) are destroyed. The quest to make smaller and smaller devices is not just a desire to have Dick Tracy wristwatches; a small device is cheaper to manufacture and less prone to defects than a large one. More importantly, faster supercomputers can be built if the time taken for a signal to go from one part of the computer to another is made as short as possible. In silicon chip technology, as much as 50 % of the available space is taken up by interconnections and wire contacts.

The fastest computers deliver 10000 billion instructions (or simple operations) per second; in jargon, 10 giga 'flops', or floating point operations per second. A next generation of computers is already needed to solve challenging problems such as forecasts of global warming trends and ozone layer spreading. Such machines will run at 'tera' flop speeds, or 100 times faster than the fastest computers of today. One way of speeding up computers is to make shorter connections between the various communicating parts (or higher 'integration'). Shorter connections and thinner dataways (presently about 50 millionth of a centimeter) will make computers run faster but hotter. Superconducting wires and interconnections can help realize the goal of shorter connections, since conventional component integration cannot be pushed much farther than today's achievements due to the difficulty in removing heat generated within the integrated chip by resistive elements.

There is another way superconducting technology can be used in electronic circuits. The processing of signals, or impulses of voltage, by the chip can be made much faster when using Josephson junctions rather than transistors. Here is how it works. Computers work using binary languages, or, in other words, instead of the 20 + letters of the alphabet (the number varies according to the language!) there are only two symbols, 0 and 1. A device working in a binary world should be able to switch from one state (say '0') to the other one ('1') by the application of a suitable signal. In a Josephson junction we can go from a current with no voltage across the junction, see Figure 7.7, to a current which has a voltage if we apply a current (which constitutes the signal) larger than the critical current. Conversely, the device remains in the state '0' if the current is smaller than the critical current. The advantage of using Josephson junctions is the greater speed with which the device processes the signal.

IBM spent many years and much money developing a computer using superconducting technology. However, it discovered that manufacturing capabilities at that time could not provide Josephson junctions in large enough numbers with the stringent specifications needed. The project was abandoned in the mid-1980s. However, Josephson junctions have been successfully employed in

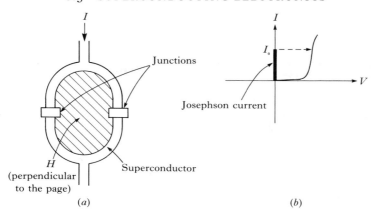

Figure 8.11(a)   SQUID employing two tunneling junctions. The magnetic field H is inside the loop. (b) I(current)–V(voltage) curve for a Josephson junction. When the current exceeds I_0, a voltage drop suddenly appears across the junction (see dashed line); this type of junction can be used as a switch.

smaller electronic circuit boards than the ones originally envisioned.

Another use of the Josephson junction is illustrated here. Suppose a voltage difference is maintained across the junction. We recall that each superconductor can be described as a giant atom; in the present case the two 'atoms' have two different energies because of the difference in the imposed electrical potentials. From Bohr's model we know that the energy of an atom can be decreased if one of its electrons is able to jump from a higher (farther from the nucleus) to a lower (closer) orbit (the orbits are quantized). In doing so, the energy difference which the electron sheds is converted into light with a frequency characteristic of the jump in energy (recall that $E = h \times \nu$, see Section 4.2). In the case of the junction, for a fixed external potential, a well defined frequency is obtained. Thus, the junction can serve as a frequency standard for instrument calibration.

Josephson junctions are used in SQUIDs (Superconducting QUantum Interference Devices) for very precise measurements of small magnetic fields (see Figure 8.11). Before we illustrate the use of SQUIDs, let us consider the following experiment. We know that a superconducting body in the shape of a ring, for example, can be characterized by a wave-function when considering its superconducting properties. Mathematicians would say that the wave-function has to be single-valued, that is, if we go around the ring we should obtain the same wave-function as at the point where we started. When a magnetic field is applied (and we suppose that it is not large enough to destroy superconductivity), calculations show that the continuity of the wave-function imposed above will be satisfied only for certain discrete values of the magnetic flux (which is equal to the magnetic field times the area of the ring). In other words, the magnetic field is quantized. The minimum value of the flux, or flux quantum or fluxoid, is 2×10^{-7} gauss $\times$ square centimeters, a very small number. Incidentally, by a careful measurement of this flux quantum it is possible to find the electric charge carried by the Cooper pair. As expected we find this charge to be the charge of two electrons.

Since the magnetic flux is proportional to the area of the ring, if the area of the ring is 1 square centimeter, very small magnetic fields, as compared to the Earth's magnetic field (0.5 gauss), can be detected. In the SQUID one or more Josephson junctions are placed in a superconducting ring. Different schemes are used to determine how much current is present in the ring; for each fluxoid which penetrates the ring there is an additional supercurrent induced and, consequently, a change in the read-out voltage of the instrument. SQUIDs are used in low temperature physics research as well as in biophysics to detect tiny changes of magnetic fields in the brain.

Most of these developments took place in the sixties and seventies; indeed once the theoretical framework was put into place in the late fifties and early sixties, technological applications followed rapidly especially after the discovery of niobium–tin alloy in 1960. A major jolt to scientists and engineers working in this field occurred in the late eighties, when completely new super-conducting materials were discovered with characteristics vastly different from the early ones, as we will see in the next two chapters.

9

High temperature superconductivity

9.1 The event

NOW THAT WE HAVE ACQUIRED some knowledge of what (low temperature) superconductivity is, we can appreciate the uproar that followed the discoveries of Georg Bednorz and Alex Müller (IBM Research Laboratories, Zurich, Switzerland) in 1986 and Paul Chu (University of Houston) and Maw-Kuen Wu (University of Alabama, Huntsville) in 1987. As we did for Onnes' discovery at the beginning of this century, let us look back at how Bednorz, Müller, Chu and Wu came to discover high temperature super-conducting materials in 1986 and 1987.

As we mentioned in the Introduction, perhaps the most celebrated moment in the history of the discovery of high temperature superconductivity was the American Physical Society Meeting in March 1987. In March of every year a few thousand scientists, most of whom work in the field of condensed matter physics (which is the study of the physical properties of solids, including superconductors) gather to present and discuss the latest results of their research. These meetings are usually quite hectic; people move in and out of meeting rooms as well as restaurants from early in the morning until well past midnight in a frenzied activity to meet new colleagues, see old friends and catch up with the latest turn and twist in the physics world. But nothing was ever as feverish as during that meeting in March 1987.

That year the meeting was held in New York City at the Hilton Hotel. By and large most of the participants were aware of the new findings of Bednorz and Müller which had appeared the year before in a respected (but, perhaps, less widely read) European physics journal: the *Zeitschrift für Physik*. In that article they reported, as the title indicates, 'Possible High T_c Superconductivity in the Ba–La–Cu–O System'. In this context, high T_c (high critical temperature) meant the breaking of the 30 K barrier, a considerable achievement if we consider that previous progress on raising T_c had proceeded at an exasperatingly slow pace. During the roughly 60 years, from Onnes' discovery of superconducting mercury wires at 4 K in 1911 to 1973 when the critical temperature of 23 K was reached for a niobium–germanium compound, the critical temperature inched up at an average rate of about 1 K every three years. Clearly, no known cause could have produced this almost regular increase of T_c over such a long period of time, and this trend was considered a statistical curiosity as well as a reliable 'old wives' tale' predictor of things to come. But after reaching 23 K, it seemed that the critical temperature was stuck at that level; indeed, about a dozen years went by before a material with a higher critical temperature was discovered by Bednorz and Müller. It is, then, understandable that scientists became as excited as they did when in the interval of just one year – from early 1986 to early 1987 – T_c went 'through the roof' from around 30 K to more than 90 K (Figure 9.1).

The word 'possible' in the title of Bednorz and Müller's article reflects both their cautious approach and the fact that diamagnetic measurements, a more telling way of confirming superconductivity, hadn't been made at that time because the necessary equipment wasn't available. Reports of 'high' T_c materials – higher than 23 K – had occasionally appeared in the literature over the years. Perhaps the most celebrated case was the observation by a group of Russian physicists in 1978 of anomalously high diamagnetism, a condition for superconductivity as we know now, in a compound made of copper and chlorine, CuCl (cuprous chloride). Critical temperatures as high as 140 K (about $-130\ ^\circ$C) were mentioned, although most of the experiments were done not

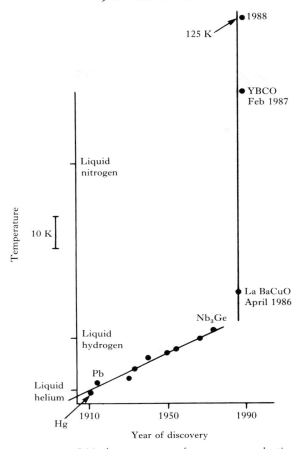

Figure 9.1   Critical temperature of new superconducting
materials versus year of discovery.

in thermal equilibrium, but during quick thermal cycling of the
samples. Bednorz and Müller knew that without an apparatus to
verify the Meissner effect (or complete exclusion of magnetic field
from the inside of the sample), their claims of superconductivity
would be greeted with skepticism. This was the reason for using
such a cautious title in their article. Fortunately superconducting
behavior was readily and convincingly confirmed in many laboratories throughout the world shortly after Bednorz and Müller's
experiments.

During the week long meeting of the American Physical Society, a special session was held (March 18) to give researchers the opportunity to present their latest findings about superconductivity. At the registration booths, clerks asked attendees whether they planned to attend this special session in order to allocate enough space and staff for the event. Reports of 90 K superconductors had already reached newspapers and curiosity was running high. The session started in the early evening and proceeded well into the next morning. Notwithstanding the size of the ballroom at the Hilton where the session was scheduled, few managed to find a place to sit; the majority stood up for hours in the main room or outside in the corridors where close-circuit TV sets had been placed. Animated conversations were held throughout the hotel. People were continually coming and going from the main ballroom seeking colleagues to share their comments with or trying to obtain precious pieces of information. Special panels of experts in 'old' low temperature superconductivity were quickly formed so that technical papers could be reviewed expeditiously for publication in the most prestigious journals, such as *Physical Review Letters*. So little was known about these compounds, their preparation methods and their physical and chemical properties that a timely publication of the latest results was deemed essential. From then on, an incredible number of Federal Express parcels and FAXes were exchanged, as it seemed that every moment was precious. And in a certain sense they were. A year later, FAX transmissions of news-breaking events in superconductivity were still so intense that the highly regarded magazine *Science* published an article about the latest finding in high temperature superconductivity with the title: 'Superconductivity: the FAX factor'; it discussed the use of just-delivered FAXes by large numbers of presenters at international superconductivity conferences.

Why was there so much excitement amidst a group of people (physicists, chemists and materials scientists) known for their restraint and even skepticism? Probably every scientist present at that American Physical Society meeting had a different answer to this question. Aside from those directly involved in research on the new phenomenon, for whom the APS meeting was just another hectic moment in the already hectic lives they had been leading for months, the majority of physicists in the audience had, perhaps, two major reasons to feel excited. One was their curiosity about this discovery, since everybody knew how slowly research on new and promising superconducting materials had progressed in the past. In particular, many were intrigued by the fact that a surprisingly high critical temperature had been found for such 'unpromising' materials as the oxides. These compounds consist of oxygen atoms bound with other elements, typically metals: they are usually electrical insulators in the non-superconducting, or normal, state, although they can become electrical conductors if small quantities of other elements are added.

Reports of superconducting oxides had appeared before (for example, in materials such as strontium titanate and lanthanum titanates, which are compounds of the metal titanium with the elements strontium, lanthanum and oxygen), but their super-conducting transition temperatures remained well below those achieved in alloys of metals. Nonetheless a small number of people were conducting research on these materials. For example, Bednorz, a crystallographer and crystal grower, and Müller, a senior scientist, IBM Fellow, and solid state physicist by training, had been working on oxides since 1983 as part of a systematic investigation of their superconducting properties (if any). The idea to use oxides stemmed from the fact that in some cases oxides of metals have a higher T_c than the pure metals (this is true for aluminum, for example). Furthermore, previous investigations on the effect of doping of superconducting oxides, that is, of introducing small amounts of another element into the compound, showed that this process was effective in raising the T_c. In one respect, their search for a new class of superconductors appeared hopeless: there were just too many ways in which elements could be mixed together (with different proportions and under different conditions) and made to react with oxygen to form oxides; without some clues, it would have been a blind and frustrating search. But their research was never mindless; the compositions of the materials they made and tested for superconductivity were the results of long searches in the scientific literature and of analyses of past failures.

Eventually their perseverance and attention to the work of their colleagues paid off. A turning point came in 1985 when Bernard Raveau and Claude Michel, two French chemists, synthesized a ceramic material containing barium, oxygen, lanthanum and copper. The Swiss scientists tried the recipe of the French chemists and these specimens lost electrical resistance at around 30 K. Soon after, this result was confirmed by Masake Takashiga of the University of Tokyo. In short order many other laboratories were able to make and test what we call now high temperature superconductors. In 1987 Bednorz and Müller were awarded the Nobel Prize in Physics for their work: this was the shortest time ever between a discovery and the award of this prestigious prize. It was the second Nobel Prize in Physics given to staff members of the IBM–Zurich laboratories within the space of a few years; the other was awarded to Benning and Rohrer for the invention of the scanning tunneling microscope (an instrument that uses electrons to image atoms on surfaces of materials).

One of the obvious reasons for the excitement of both scientists working in the field of superconductivity and other physicists as well was that now there seemed to be a real possibility of creating 'room temperature' superconductors, a possibility many had dismissed as totally unrealistic before. The other, more ethereal reason for this unusual feeling of expectation was that these compounds might display a mechanism of superconductivity different from the one suggested by the BCS theory. In this very successful theory the interaction between electrons in the lattice is mediated by the lattice itself and this interaction, usually repulsive for two isolated electrons, becomes attractive when placed in a lattice with appropriate properties (the electrons are said to form Cooper pairs). The fact that now there was a *big* problem waiting to be solved (a new mechanism for superconductivity? new interactions? exotic quasi-particles?) was intoxicating to many.

But some people who showed up at that meeting were motivated by other considerations. The relative simplicity of making these compounds (although, at that time, the mixing and other exotic procedures required to obtain successful samples resembled, to many physicists, alchemy more than anything else) and the demand to obtain a large quantity of information about the chemical and physical properties of these materials, prompted many researchers in diverse fields of expertise to try to 'jump into the game' and start working with these new compounds. Scientists at universities and at major industrial laboratories with a background (or even a friend with a background!) in chemistry or materials science were avidly

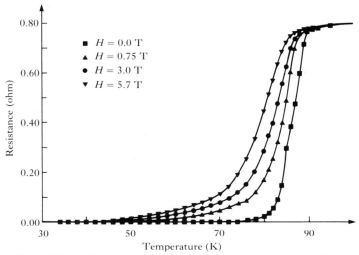

Figure 9.2   Resistance vs. temperature for YBCO (yttrium–barium–copper–oxygen). Data are taken at different magnetic fields (in tesla). From: *Physical Review Letter*, **58**, 908 (1987). (Courtesy of Paul Chu; copyright: American Physical Society.)

sought, since the chemicals and the mixing and firing procedures familiar to ceramicists and inorganic chemists were not familiar to most of the physicists.

Following the announcement of the work of Bednorz and Müller and before the American Physical Society meeting of 1987, a few groups throughout the world started to work on that class of compounds, changing the proportions with which the various elements were mixed together or replacing some elements with others that had slightly different characteristics. The most spectacular discovery was made by Paul Chu of the University of Houston and his collaborator Maw-Kuen Wu of the University of Alabama in Huntsville. In a series of experiments carried out at the end of 1986 and the beginning of 1987, just before the March meeting of the American Physical Society, they not only succeeded in repeating the results of Bednorz and Müller, but obtained samples which showed the onset of superconductivity at temperatures above 90 K, a value spectacularly high (Figure 9.2). For the first time there was the possibility that devices cooled with the rather inexpensive liquid nitrogen (which boils at 77 K) could exploit superconductivity.

The key to these discoveries was the substitution of the element lanthanum by yttrium, the latter an element belonging to the class of 'rare earths'. This class of elements has some peculiar characteristics but is by no means 'rare' in nature. Yet the rush to substitute elements for those in the original compounds, put the rare earths temporarily in short supply in many laboratories. At the meeting one could hear many physicists asking colleagues about the availability and purity of the most sought-after chemicals. Everybody realized that speed was important and that soon hundreds of laboratories would be working on these materials; it was essential to obtain the raw compounds as soon as possible, mix them, cook and characterize the samples, and perform measurements that nobody else had yet done. If successful, one would obtain almost instant recognition; he or she would be able to attract the best graduate students and collaborators and would have a better shot at securing funding for the laboratory, always the principal headache of any scientist with an active laboratory (at least in the USA).

In this frenzy to change chemicals in order to hit up on the right combinations of elements which would yield the Holy Grail (superconductivity at room temperature) some rational guiding principles were used. In the last century, the Russian scientist Mendelev produced a table, now called the Periodic Table of the Elements, in which the different elements (atoms with a different number of electrons) were neatly organized into columns and rows, or a table, according to their electronic characteristics (the specific way electrons were arranged around the nucleus of each elemental atom). This organizational chart was, and is, very useful, since these arrangements of electrons can suggest the propensity of a given atom to cede or acquire an extra electron from another element (atom) in order to form a more complex entity. Using this chemists could make a reasonable guess as to the properties of a compound formed with different elements of the Periodic Table even before mixing the actual materials.

In Houston, Paul Chu, who had been working for years on superconducting compounds, found that on applying pressure, that is, squeezing the sample, the transition temperature went up. This meant that atoms, when subjected to pressure, shortened their relative distances apart, producing a denser, more compact solid. Even though the average separation between atoms changed only by a very small fraction, a remarkable thing happened: their electric properties changed drastically. Chu and his colleagues were right; by changing the atomic structure, the temperature of the onset (or 'beginning') of superconductivity could be raised. However, it was not clear why atoms being closer together should have such an effect on superconductivity. More work was needed; besides, it would have been impractical to use high temperature superconducting materials only when an external, and considerable, pressure was applied to them. Nonetheless, it was a very promising starting point. Was it possible to mimic this change in the position of the atoms without applying an external pressure? For example, if some of the chemical elements present in the compound were replaced with others found by the skillful use of the Periodic Table, could one achieve the same effect as that obtained when using external pressure? One could try to squeeze 'larger' atoms into the existing lattice, thus obtaining a pressure using 'internal' means. Paul Chu and his collaborators were not the only ones following this path.

After the announcement that 90 K superconductivity had been attained by Chu's group a race to find an even better superconductor began. Once it became clear that introducing a certain type of impurity into the lattice raised the critical temperature, workers in many laboratories throughout the world started substituting yttrium and also barium (two key elements in YBCO) with all sorts of elements. When asked what kind of substitutions they were trying, researchers often responded: 'We went through the Periodic Table'. This sweeping statement indicated both that they had been working hard and systematically to obtain better materials, and that they were not going to give away the key to their search so easily.

Later that year (1987), when several other groups were working on the new superconducting materials, the press reported ever higher critical temperatures in a dazzling succession. There were claims that materials had been discovered in which the onset of superconductivity occurred at a temperature as high as room temperature.†

---

† In some of these new materials, the transition to the superconducting state was not sharp, but gradual (see, for example, Figure 9.2, the line with square symbols); most of

Some of the findings reported (with increasing frequency) in the scientific and in the lay press were not corroborated by experiments in other laboratories and soon faded into oblivion. Although the critical temperature jumped from 23 to 30 and then to 90 K in less than a year, it stopped increasing in the following years. The newest compounds which reproducibly display the highest critical temperature (125 K) were discovered by several laboratories in 1988; they contain the elements thallium or bismuth in addition to oxygen and copper. Since then, no material with a higher critical temperature has been discovered.

In reading press coverage from those days one gets the impression that a blind race was on to obtain The Prize, to fabricate a compound with the highest critical temperature. There was, obviously, tremendous pressure to find and patent the best compounds; however, a parallel and, ultimately, even more exciting effort began – to measure the properties of these new superconductors and find out whether they were unusual BCS-type superconductors or whether they used entirely new mechanisms which had to be proposed and verified. Theorists were as busy trying to make sense of the large amount of data obtained as experimentalists were changing chemical elements in compounds. But before we look into the proposed mechanisms, we have to examine some key properties of these new materials.

9.2   Characteristics of the new superconductors

For the purpose of discussing high temperature materials, we divide them into three classes. The first class is the one that Bednorz and Müller found; these materials contain lanthanum, barium, copper and oxygen and exhibit a critical temperature between 30 and 40 K. The second is the so-called 1–2–3 compound that Chu and collaborators discovered (T_c of about 90 K); it contains yttrium, a rare earth element, rather than lanthanum and has the following proportions: 1Y 2Ba 3Cu 7O, (1 part yttrium, 2 parts barium, 3 parts copper, 7 parts oxygen) or 1–2–3 for short. (The class of materials in which the proportion of the various

the claims referred to the beginning – or onset – of superconductivity. Thus, this should be considered as an upper limit.

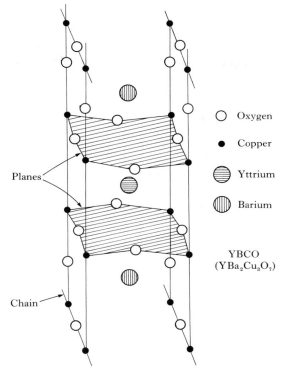

Planes

Chain

Oxygen

Copper

Yttrium

Barium

YBCO
($YBa_2Cu_3O_7$)

Figure 9.3 Arrangements of atoms in a crystal of $YBa_2Cu_3O_7$; superconducting YBCO has some oxygen atoms missing.

elements changes slightly with respect to the 1–2–3 compound mentioned above is also called YBCO). The structure of this compound, that is, the relative positions of the atoms in a periodic arrangement in a crystal, is called a perovskite, a term often mentioned even in newspaper accounts, and is quite complex as can be inferred from Figure 9.3. Here it suffices to say that there are two peculiarities deemed important for superconductivity in this type of structure which are: first, oxygen and copper atoms are bound together in alternating layers containing chains and planes; second, these materials show strong anisotropy in many of their properties; for example, electrical conductivity has a very different value when measured parallel to the oxygen–copper planes than when measured perpendicular to them. Essentially these can be considered as layered compounds in which the 'action' takes place within the oxygen–copper planes. We will see in the next chapter that this anisotropy has important consequences for the fabrication and utilization of these materials.

As we said above, following Chu's discovery, his group and others throughout the world methodically explored the effect on T_c of varying the proportions of the elements forming the compound as well as by substituting other rare earth elements for yttrium. Several different superconducting materials in the YBCO class have been realized in this way, although the highest transition temperatures for this class of materials remains in the 90 K range.

The third class of materials, discovered in 1988 by teams at the University of Arkansas and the National Metal Research Institute in Tsukuba, Japan, is the one with the highest critical temperature achieved thus far, about 125 K. These materials do not contain rare earths; one is a compound of bismuth, strontium, calcium, copper and oxygen while another has thallium in it instead of bismuth and barium instead of strontium. They have a structure somewhat similar to that of the 1–2–3 materials but they are not so neatly ordered; they lack oxygen–copper chains, although they still have oxygen–copper planes. Thus, it does seem that these oxygen–copper planes are very important in creating high temperature superconductivity. This is the characteristic property toward which most research has been directed by theorists and experimentalists.

To many scientists and lay people it must seem odd that after the first materials were produced in Bednorz's, Müller's and Chu's laboratories, so many other superoxides were discovered in such a short period of time. Although initially it was mostly intuition which guided scientists in the search of the substitutional element that would give the material an even higher critical temperature than before, latterly a better understanding has been reached of the relative roles played by the various constituents of the compound.

Is there a set of characteristic physical and chemical properties which is shared by *all* the high T_c materials? The answer is that such common properties haven't been found yet. Obviously, it would be enormously important to discover some key features shared by every high T_c material. The reader might recall what happened during the search for a mechanism for low-temperature superconductivity; after a very long and tortuous hunt scientists finally hit on the right clue, that is, the interaction between electrons and lattice vibrations ('phonons'). The discovery of the 'isotope effect' (see Chapter 7) gave theoreticians the assurance that their calculations of the interaction between electrons and phonons had a key part in superconductivity. One might speculate that something similar might occur for high temperature super-conductivity. Once the key elements are singled out, it would be 'just' a matter of time before an explanation of high temperature superconductivity is proposed.

With these remarks in mind, we will survey those properties which strike scientists as potentially critical in a successful explanation of high T_c superconductivity. It should also be noted that we are describing a field which is evolving very rapidly; thus, some of the features of superconductors we are going to illustrate might become irrelevant in the very near future, while others will come into prominence. In these layered materials, super-conducting properties are believed to be confined to the copper–oxygen planes separated by yttrium (or another rare earth element), while the other layers contain the remaining oxygen, copper and barium atoms. If the former (those with copper–oxygen planes) are called the conducting layers, the latter are called the charge reservoir layers. These layers have the function of providing positive or negative charges for the conducting layers. Depending on the size and propensity of each rare earth or metal atom to capture or yield electrons, the material acquires different electrical and structural characteristics.

One of the most intriguing characteristics of these materials is that, for a given compound, superconducting behavior persists over a range of oxygen compositions. Specifically, it has been found that not all of the oxygen atoms residing in the charge reservoir are in their correct, ideal sites; rather, some of the oxygen atoms are either missing or pushed out of position. It is now recognized that superconducting properties depend on these defects (or departures from an ideal crystal). Variations in the concentration of defects, such as missing oxygen atoms or the presence of some other elements between the copper–oxygen planes, in the conducting layers can have even more remarkable effects, such as the destruction of the superconducting state altogether. Unfortunately, this added level of complexity makes theoretical calculations of many properties of these materials particularly difficult to carry out.

Although much progress has been made in understanding the major structural and electrical properties of these new super-conductors, little is known about how the superconducting state is realized in these materials. As we said earlier, one of the reasons there was so much excitement about the discovery of 90 K superconductors was the belief that these new superconductors might have different mechanisms for superconductivity than the one given by the well proven BCS theory. If indeed these materials do not obey the BCS theory based on the pairing of electrons via the 'mattress effect', what is the new mechanism? Are there clues, such as peculiar physical properties in these materials, which can lead us to the discovery of this new mechanism as occurred for the low temperature superconductors?

As is to be expected, the initial flurry of activities was concentrated on measuring key quantities to explore these issues. For example, as we saw in a previous chapter, infrared absorption measurements could be used to measure the energy gap in a superconductor separating the Fermi sea (where the electrons live at $T = 0$ K bound in Cooper pairs) from the excited states (where pairs are broken up); or a flux quantization experiment could be done to see whether the charge carriers couple in pairs as in the BCS theory and to determine the amount of charge carried; or other experiments could be carried out to estimate the spatial extent of these pairs (that is, the spatial 'coherence' of the wave-function, see Chapter 7). These and many other experiments were indeed performed in laboratories all over the world, but not only did they fail sometimes to provide unequivocal answers, they often raised more questions than they had attempted to answer. Although a comprehensive understanding of the new super-conductors has not been reached, nonetheless, there are a few results which are agreed upon by a number of experts in this field. As the samples become better characterized and more complete theoretical models are introduced, we can expect to obtain better and more easily interpretable results.

Undoubtedly there are a number of similarities between the conventional and the new superconductors. For example, in both cases the charge carriers are bound in pairs, as determined by the experiments using magnetic fields, illustrated in the previous chapter. These new pairs differ from BCS pairs in one respect at least: the distance between the charge carriers of each pair in the new superconductors is much shorter, by a factor of around 100. Furthermore, the isotope effect, which was seminal in the formulation of BCS theory, is almost absent in the new superconductors. This means that changing the nuclear mass of the oxygen ions in the lattice slightly has little effect on the superconductivity; this seems to indicate that the role of phonons in the new materials is non-existent or, at least, different than in BCS superconductors.

There is still an energy gap between the filled energy states of the electrons and the empty ones; as we illustrated earlier this is an important characteristic of the 'old' superconductors. Since there is an energy gap one can still think of the excitations of the superconductors in terms of quasi-particles (see Chapter 7), although it is not clear yet what the nature of these quasiparticles is. In BCS theory, each pair that falls apart produces two quasi-particles; an electron is pushed above the Fermi level and a vacancy, or 'hole', is left in the Fermi 'sea'. For the new superconductors, other types of quasi-particles have been proposed, although no conclusive experiment has yet provided evidence for the existence of one type of excitation over another.

Another similarity between the old and new superconductors is their behavior in the presence of an external magnetic field; the new superconductors can be considered essentially as Type II superconductors in which the external magnetic field is allowed to penetrate into the sample through bundles of magnetic field lines separated by superconducting regions. The typical critical magnetic field of the new materials is surprisingly high; this is a welcome piece of news for engineers, since these materials can be used in superconducting magnets to generate much higher magnetic fields and carry much higher currents before they revert to the normal state. There is, however, a problem: it appears that the current densities (electric current divided by cross-sectional area) which can be sustained in the new superconductors are much smaller than in conventional superconducting materials. It is thought that this inability to carry large currents is due to the lack of 'pinning' of the fluxoids, which are the bundles where the magnetic field is concentrated in the superconductor. If these bundles were free to move around the crystal when pushed by the passage of an electrical current, as is the case when there is no pinning, dissipation of energy would occur and a degradation of superconducting behavior ensue, eventually returning the material to the normal state. There are reasons to believe that these fluxoids behave less like a neat array of parallel flux – or vortex – lines (a simplified view of the behavior of the old Type II superconductors) and more like a bunch of tangled spaghetti, called a vortex fluid. At lower temperatures, it is speculated that a 'glass' of vortex lines is formed (Figure 9.4). Understanding how this tangle of spaghetti lines is formed and moves is an active research topic.

The materials which we have been discussing so far are the ones which were found to be superconducting following the discoveries of Bednorz, Müller, Chu and others. As illustrated above, these materials might have very different properties, but they share the general feature of being layered compounds. In these solids layers of oxygen and copper atoms are stacked on top of each other; depending on the specific material under consideration, other layers, chains of atoms or even isolated atoms are inserted between the oxide layers.

There is, however, another type of superconductor, which has been discovered in the last two years, with transition temperature now exceeding 30 K. Were it not for the discovery of the superconducting oxides a few years earlier, these materials would have commanded the attention of newsmedia as the cuprous oxides did. Instead, their discovery has been reported almost exclusively in scientific journals.

These materials contain a new form of the element carbon. Carbon atoms are known to arrange themselves in a lattice in two ways, forming two solids with very different physical and chemical characteristics, diamond and graphite. Now chemists have discovered that they can arrange carbon atoms in a new and surprising way. They assemble 60 carbon atoms at the vertices of a structure which resembles the geodesic dome of the famous architect Buckminster Fuller. For this reason, these assemblies of atoms (or molecules) have been nicknamed 'buckminster fullerenes' or, for short, 'buckyballs'. The name 'ball' has stuck to these molecules

(a)

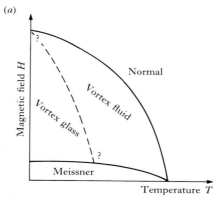

(b)

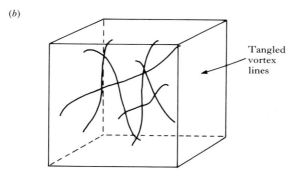

Figure 9.4(a)   Tentative 'phase diagram' for high temperature superconductors. The superconductor can be in different phases or states, depending on the values of temperature and magnetic field strength. Complete expulsion of an external magnetic field occurs in the region labeled 'Meissner'. (b) Schematic representation of wandering flux lines.

because the atoms are arranged in a cage with the shape of a soccer ball (the dimension of this ball being about 10 angstrom). A more scientific notation for this molecule is C_{60}, where C is the symbol of carbon and 60 indicates the number of carbon atoms. When these balls are put together and linked by metal atoms such as potassium, they form a compound which becomes superconducting around 30 K. Little is known about these superconductors; at present it seems unlikely that the mechanism of superconductivity in potassium-doped buckyballs would be similar to the one in the cuprous oxides.

9.3   Mechanisms of superconductivity

How does superconductivity occur in these new materials? For the old ones the key to the explanation of superconductivity was the attractive interaction between two electrons, although, as we well know, electrons in empty space repel each other; for super-conductors this attraction can take place, under appropriate circumstances, with the help of the ions of the solid. There is a parameter in the BCS theory which is related to the strength of the interaction between an electron and vibrations (phonons) in the lattice. Simplifying a bit, the stronger the interaction, or coupling, the higher the critical temperature. However, if BCS theory is to explain high temperature superconductivity, theorists have calcu-lated that this electron–phonon coupling parameter cannot give rise to a superconducting transition temperature much higher than 30 K. At these temperatures it is believed that the large vibrations of the lattice would disrupt the role of the lattice in providing the attraction between two electrons (recall the 'mattress' effect discussed in a previous chapter). More and more scientists are now convinced that another mechanism, or mechanisms, must be at work, but what? If history is going to repeat itself we are in for a long wait. Recall that it took 46 years from Onnes' discovery to the formulation of the BCS theory. Of course there is no reason why this should be so for the high temperature superconductors, and, in fact, we are now much better equipped with experimental and theoretical tools to meet this new and unexpected challenge.

Indeed several alternative mechanisms have been proposed. Essentially what theorists are doing is trying to find other types of mediators between the electrons rather than the lattice envisioned by the BCS theory. As is true for phonons in the BCS theory, these mediators should transform a repulsive force between two elec-trons into an attractive one. But before discussing these new ideas we should make an observation about the magnetic behavior of the new materials.

Generally speaking, magnetism and superconductivity do not mix. If a solid is magnetic, for example, ferromagnetic like a piece

(a)

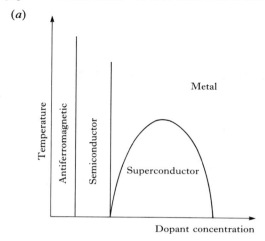

(b)

Antiferromagnetic chain

Figure 9.5(a)   Temperature and dopant concentration for which different phases (states) of copper oxides are stable. (b) Schematic representation of an antiferromagnetic chain.

of magnetized iron or permanent magnets, the chances of it being a superconductor at the same time are very slim. The explanation for this is that magnetic ions in the solid (ions which produce a magnetic field of their own) are very good scatterers of the Cooper pairs and therefore can destroy the long range coherence which is the essence of a superconductor. It is interesting to note that many of the high temperature superconductors display a rich variety of magnetic and electrical behavior depending on the temperature and relative composition of the elements of the compound (Figure 9.5). Superconductivity is achieved only at low temperatures (generally 90 K and less), and only for a certain range of electron concentration in the copper–oxygen planes (this concentration is dictated by the presence of 'dopants' or impurities). For other values of temperature and concentration, the material can become metallic, semiconductor, or magnetic. To be more specific, several of the new superconducting materials are antiferromagnetic which means that the atoms of the solid carry magnetic moments.† In an antiferromagnet, these moments are arranged so as to get a zero averaged magnetic moment. For example, if I had an antiferromagnetic chain of atoms, their moments would be arranged to produce an alternation of moments in one direction (or 'up') with those in the opposite direction (or 'down'). Instead, in a ferromagnetic chain all the magnetic moments are 'up', and a net magnetization is observed.

Electrons too carry a magnetic moment, in jargon called 'spin' – the same spin that we introduced when we discussed Fermi statistics in an earlier chapter. Theoretical models have been developed in which the attractive force required for pairing of electrons is not due to the lattice (or lattice vibrations) but to magnetic forces associated with the spins of the electrons. Phil Anderson of Princeton University (Figure 9.6), who has made important contributions to the theory of 'old' superconductors, hypothesized that the spins of the electrons form a 'liquid' which, under appropriate circumstances, shows superconducting properties. Other approaches, such as the one laid out by William Goddard III of the California Institute of Technology (Figure 9.7), rely more on the chemical properties of the various elements in the compound. In this case, correlations between charges in the oxygen atoms are claimed to take a key part in the superconductivity of the oxides. Other theorists think that the pairs are not formed when the solid becomes a superconductor, but are already present in the normal state; however, in the latter case, the motion of the pairs is not correlated (recall that a superconductor is essentially the realization of a giant, coherent atom). At the transition temperature all the scattered pairs get in step or 'condense' into a new coherent state.

So far no comprehensive theory has emerged which can explain all the superconducting properties observed in the new materials. The theories proposed differ greatly; some are more phenomenological, some extensions of BCS and others quite radically new. In any case they haven't been developed to the fullest extent and

† Magnetic moments can be visualized as tiny dipoles; the needle of a compass is a good example on a macroscopic scale.

Figure 9.6   Philip W. Anderson. He won a Nobel Prize in Physics for his theoretical investigations of properties of disordered systems. (Courtesy of the *New York Times*.)

their predictive powers, a key test of the usefulness of theoretical models, have not been proven yet due to the complexity of the calculations necessary to compare theoretical results with experimental findings. The title of an article ('High Temperature Superconductivity: Where is the Beef?'), which recently appeared in the well regarded magazine *Science*, seems to echo the opinions

Figure 9.7   William Goddard III. (Courtesy of the *New York Times*.)

of some scientists who think that many theories have been proposed but few results have been obtained. A few well known theorists angrily challenged the conclusions of that article. But what do the experts say?

'We don't know the mechanism of high temperature super-conductivity yet': this was the opening remark of W. A. Little, who presided over the 'Symposium on the History of Physics' in a special session of the 1991 March Meeting of the American Physical Society devoted to the discovery of high temperature superconductors.

Phil Anderson, on another occasion, said: 'The consensus is that there is absolutely no consensus on the theory of high T_c superconductivity'.

Actually these remarks should perhaps be interpreted as a recognition of the challenge theorists face and of the multitude of approaches followed, including some old ones. For example, some claim that the superconducting properties of the new materials can be explained within the framework of the existing theories (BCS and interaction between electrons mediated by the lattice) which explain rather satisfactorily superconductivity in most metals, but not in other more exotic materials, such as organic superconductors (superconductors made with large and complex molecules with transition temperatures around 11 K). Anderson, who proposed a few years ago a mechanism of high temperature superconductivity based on exchanges of new particles (not yet detected), believes that the hard part of the problem is not so much to come up with a reasonable mechanism, but to explain coherently the unusual properties of these materials both above and below the transition temperature. Incidentally explaining the electrical characteristics above T_c of many (if not most) of the new high T_c materials has proven very challenging. For example, the resistivity is found to increase with temperature almost linearly instead of the much steeper rise found in metals (ρ increases with T^5). Although Bob Schrieffer, the 'S' in the BCS theory, is pursuing a different path than Anderson's, he too believes that to find new ways to handle these new properties from a theoretical standpoint is more important than to find a specific mechanism which mediates the interaction between the electrons.

Unfortunately, we have to stop here, since these and other mechanisms which have been proposed for high temperature superconductivity require an advanced background in physics to be understood properly. Instead, in the next chapter, we will examine the progress which has been made in implementing the new superconductors in devices and we will illustrate the difficulties which prevent the rapid technical realization and commercialization of high temperature superconductors.

10

Technological applications of the new materials

10.1　The food chain

AT THE FAMOUS MEETING of the American Physical Society in March 1987, a sample ribbon of the new high temperature superconducting material was shown. To many it seemed that the implementation of high temperature superconductivity was just around the corner.

Now that the breakthrough in high temperature superconductivity is a few years behind us, the general public and experts alike ask whether this scientific discovery has also been a technological breakthrough. As we have seen in a previous chapter, the discovery of Type II superconducting materials was quickly followed by the commercialization of superconducting magnets; similarly, the discovery of the Josephson effect and the successful fabrication of Josephson junctions opened up the field of precision measurements and SQUID magnetometry. What technological fruits are high temperature superconductor materials going to yield? Are the new materials going to bring about a technological revolution?

It is easy to see how the media sometimes play on the fancy of the public by envisioning a near future run by gadgets that exploit in every possible way the 'magic' properties of high temperature (or

perhaps even room temperature) superconductors. Superconducting power lines, magnetic levitated trains, superefficient electric
cars, non-invasive medical imaging, superconducting ships, superconducting supercomputers and bigger high energy particle
colliders (however, the superconducting supercollider under construction in Texas will use low temperature superconducting
materials – see Chapter 8): these are just a few of the possible
applications presented to readers and TV viewers. But is it
probable or even possible to have superconductivity working in
these gadgets in the near (10–20 years) future?

There are plenty of examples to show the fact that technological
wizardy is not always synonymous with commercial success (the
much talked about 'paperless office' is still a few years away). How,
indeed, do scientific achievements or breakthroughs get translated
into commercial products? How long does it take? Let us see how
much progress has been made in implementing this discovery for
commercial use.

It is encouraging to note that at meetings of scientists working on
the new superconducting materials a good deal of time is now
devoted to technological aspects of these materials. Another sign
that commercial products are not far away in the future is the
tremendous number of companies, some old, some new (including
many new 'start-ups'), which are devoting a great number of
technical and financial resources to the development and commercialization of high temperature superconductors.

As Dr John Rowell of the newly formed company, Conductus
Inc., put it in his plenary lecture at the American Vacuum Society
Annual Meeting, held in Toronto in 1990, one has to consider the
food chain, that is how a scientific discovery is translated into a
product. This food chain requires many steps; it is analogous to
what happens in the animal world when small fish are eaten by
bigger fish, who are eaten by still bigger fish, etc. After a newly
designed material has left the research laboratory, it is applied as
the component of a subsystem, for example as an electronic
component of a circuit board. As engineers learn more about the
material, new and larger applications, perhaps subunits (such as
the integrated electronic chip, for a bigger system) are envisioned.
Finally, when the technology of this material is mature, a whole

system such as a computer or a scientific instrument is designed around its properties.

A few little devices which use high temperature superconductors have already entered the commercial market. Exploiting the sharp drop of electrical resistance in a superconductor, one of these devices has been used to measure, for example, the amount of liquid nitrogen in a vessel. As the level of liquid nitrogen inside the vessel drops, the output of the superconducting device changes, since more and more parts of the vessel are at a temperature considerably higher than that of the liquid. There are other ways to measure the level of liquid nitrogen, and it is not clear whether using a superconducting 'dip-stick' device has tremendous advantages over more conventional and well proven methods, but this can be considered one of the first examples of commercialization of a small product the operation of which is based on one of the properties of the new superconductors. Another examples is an infrared detector, called a bolometer, which is sensitive to electromagnetic waves of length longer than visible light, a type of radiation we feel as 'heat'. In this case electromagnetic radiation, coming from a star or an aircraft, produces enough heat in the device to cause it to go from the superconducting to the normal state. This abrupt change in resistance can be measured and the resulting signal can be amplified and processed. There have been bolometers operating at low temperature (liquid helium temperature) for quite some time. The high temperature bolometers could find many other applications where cooling to cryogenic temperatures represents a problem.

The next use of the new materials might be to have elements in an electric circuit exploit superconducting properties, such as reduced resistance in dataways or the fast switching capabilities of Josephson junction based devices. SQUIDs, which use Josephson junctions made of high T_c materials have recently been built at the prototype level. We haven't yet reached the point when these new materials can compete with their low temperature counterparts. Microwave devices using high temperature superconducting materials have been already made. The low loss and low power consumption make these devices particularly suited for space-based applications (in communication satellites, etc.). It is expected that a few finished electronic products based on the new superconducting materials will reach the market in one or two years. Presently, research and development of high temperature superconducting materials and devices is quite strong in the USA. The USA still holds 60–70 % of new patents related to applications of high temperature superconductivity. If one compares what silicon has meant to the electronics industry to the potential of superconducting materials, one can envision that one day not far in the future there will be several diverse applications for superconducting materials, each of them exploiting one of the many different properties. As David Larbalestier of the Applied Superconductivity Centre at the University of Wisconsin in Madison put it recently: 'After the stone, bronze, iron, steel, and semiconductor ages, are we going to see a superconducting age?'

10.2 Technical issues in the new materials

What are the problems confronting the engineers trying to bring out products using high temperature superconducting materials? Some of the difficulties are similar to the ones encountered using the 'old' superconducting materials, while others are due to the fact that a number of the physical and chemical properties of these materials are still poorly understood. One problem is the difficulty in obtaining materials, both old and new, which not only have good superconducting properties, such as a high transition temperature and good current carrying capabilities, but also other characteristics important for producing a saleable device, such as mechanical strength, ductility (how easily they can be shaped), and low processing costs. At the present time the 1–2–3 superconductors seem to represent the best compromise between superconducting properties, such as reasonably high critical temperature, and ease of fabrication and processing. However, recent developments make bismuth containing superconductors – called BISCCO – an appealing choice in many applications.

Some of the difficulties that materials scientists are encountering in trying to improve the physical properties of the superconducting oxides, is that these are truly metastable compounds. In other words, these materials are not in a state of equilibrium at ordinary atmospheric pressure and room temperature, but evolve very slowly into other more stable structures, as does diamond, a gem that, given enough time, will evolve into graphite, a more stable (and less precious!) structure of carbon. Although in most cases this evolution is exceedingly slow, and thus not a concern in terms of the stability of the material for ordinary uses (people wearing diamond rings shouldn't worry and can pass on their gems to many generations to come), on the other hand this metastability causes difficulties in synthesizing these compounds.

These difficulties notwithstanding, engineers hope to exploit these materials for large scale applications, such as cables for electric power lines, medical resonance imaging, maglev trains, etc., as illustrated in Chapter 8. When reliable, low cost electromagnets, electric motors and power cables can be produced from high temperature superconductors, then it is reasonable to assume that a much more widespread use of those technologies will follow. As with the low temperature superconductors, let us proceed to see what are the obstacles impeding the immediate use of high temperature superconducting materials in large scale projects.

The new high temperature superconductors are remarkably resistant to the penetration of magnetic fields. Like low temperature Type II superconductors, the new materials allow the magnetic field to penetrate the bulk in bundles (magnetic flux bundles). We recall that these 'bundles' can be visualized as small regions in which the superconductor reverts to the normal state and where the magnetic field is concentrated. These regions are surrounded by superconducting material. If an electrical current is applied, the current will flow into the superconducting regions – since in those regions it doesn't encounter any resistance. However, if bundles of magnetic field are present, the current will try to push these bundles aside and in doing so heat will be dissipated. Effectively the current finds a certain electrical resistance to its passage. Heat dissipation can be prevented, as with low temperature Type II superconductors, if the flux bundles are held (pinned) in place, for example by some impurity or defect of the crystal.

To be sure, before engineers found a way to fabricate multifilamentary superconductors of niobium–titanium and niobium–tin (the most widely used low temperature superconducting materials for high field superconducting magnets), similar problems dogged engineering research in the low temperature superconductors.

Unlike high temperature superconductors, niobium compounds are not granular and show good flux pinning characteristics. Granular superconductors have a checkerboard pattern of regions of strong superconductivity (very resistant to the presentation of magnetic fields) separated by interfacial regions which are weakly superconducting. Because of these weak links, the overall super-conducting properties of the material are seriously degraded in comparison with a homogeneous material. To be more specific, this degradation is due to the fact that, in practice, most bulk materials, as they come out of a production line, are not single crystals with the atoms neatly arranged over long distances. Single crystals of YBCO have been made (Figure 10.1), but they are extremely costly to produce. They are useful for studying the physical and chemical properties of these materials, since disorder doesn't scramble the information one wishes to obtain.

Ordinary pieces of matter, such as a steel pipe or a piece of wire, are made of many microscopic crystals, which are oriented randomly one from the other most of the time. Thus, in a steel pipe as in a low temperature superconductor, crystal planes belonging to adjacent crystallites with different tilts often meet at a steep angle, see Figure 10.2. The planes belonging to differently oriented small crystals are called grains, and the interfacial region between grains is called a grain boundary. Because low temperature superconductors are generally isotropic and the coherence length (which is the spatial dimension of the Cooper pair) is rather large compared to atomic scales, these defects do not constitute much of a problem. Unfortunately, because the coherence length of the new superconductors is much smaller than those of the low temperature ones, their superconducting properties are more sensitive to granularity on a much smaller length scale. This makes the detection and correction of a granular pattern much more difficult to achieve in practice because we don't yet have good tools to use on such a small length scale. Furthermore, because 1–2–3 and related materials are rather anisotropic, it is important to make a sample in such a way that the oxygen–copper planes, which are the most important for electrical conduction, are roughly parallel to each other in order to assure continuity of electrical conduction from one grain to the next.

As to the pinning problem, it turns out that niobium-based

superconductors can be appropriately modified so that they exhibit strong pinning characteristics. Unfortunately, the new super-conductors show very weak pinning capability and scientists haven't yet found a way to improve it. The result of the presence of granularity and weak flux pinning is that much weaker electric

(a)

(b)

For legend see facing page

Figure 10.1(a)   Micrograph of single crystals of YBa$_2$Cu$_3$O$_{7-x}$. (Courtesy of D. L. Kaiser and F. W. Gayle, National Institute of Science and Technology.) (b) Micrograph of a twinned structure (two single crystals back-to-back) of YBa$_2$Cu$_3$O$_{7-x}$. (Courtesy of L. C. Smith, F. W. Gayle and D. L. Kaiser, NIST.) (c) Micrograph of gadolinium-123. (Courtesy of M. W. Davidson.)

current densities can be sustained than an ideal high temperature superconducting material would warrant. A considerable amount of effort is being made to understand how the structure of these materials is responsible for the lack of pinning which translates into low critical current densities.

The discussion above centered on bulk material. A brighter outlook emerges for other types of high temperature super-conductors. When speaking of critical current density perform-ance, we must distinguish between bulk high temperature super-conductors and their thin film counterparts.

Thin films constitute an important component of many products. A thin film is a deposit of a certain material, not necessarily a superconducting one, on top of a substrate (top panel of Figure 7.6). The thickness might vary from just one atomic layer to hundreds of layers (still thinner than a human hair), depending on the application. Thin films are used in very diverse ways, ranging from antireflective coatings for photographic lenses for cameras, to patterned layers representing transistors and dataways in silicon-based computer chips, to anticorrosive coatings in pipes. The good news is that high temperature superconducting thin films have generally much better critical current capabilities than the same materials in bulk form, although some other properties, such as the transition temperature, are sometimes worse than those of bulk materials with the same composition. Current densities approaching the ones desired in many applications (10^5–10^6 amperes per square centimeter) have been achieved in thin films. At present, a high current carrying capability is more desirable than relatively small changes or depressions in the transition temperature and, thus, many companies are working on high temperature superconducting thin films.

One of the aims in superconductivity thin film research is to integrate a device, made with the new superconducting materials, with electronic chips which are usually made of semiconductor materials such as silicon, germanium or gallium arsenide. However, in order to obtain a film of 1–2–3 oxide which is superconducting, one needs to heat the thin film at high temperatures (several hundred degrees centigrade). Such treatments effectively preclude the use of this material in complex chip architectures where interdiffusion of atoms from one part of the chip to another might occur at high temperature, thus degrading the performance or even destroying the chip itself. These temperatures can also seriously degrade the structural and electronic properties of the underlying support which is usually made with one of the semiconductors mentioned above. A great deal of effort is being directed toward producing and processing superconducting thin films at 'low' temperatures (a couple of hundred degrees centigrade or less). Perhaps one of the most impressive stories in the

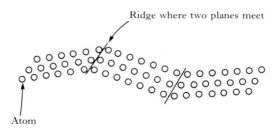

Figure 10.2   Crystal planes meet at an angle, creating ridges or boundaries.

development of high temperature superconducting materials has been the rapid progress made using thin films, both in terms of electrical characteristics (high current density capacities) and fabrication and characterization methods. A piece of positive news is that several very different procedures for making thin films yield comparable results, indicating that fabrication methods are, apparently, not very critical. Finally, we should mention that chemical stability and non-reactivity with the substrate or supporting material is also important if electronic chips have to be built. Some of the new materials are quite chemically unstable; they lose oxygen, for example, or they react with the substrate even when below temperatures reached during deposition and processing.

Finally, it is worth reporting an interesting point that has been made by many experts in superconductor technology: that is, the quest for superconductors with high critical temperatures (approaching room temperature) shouldn't become an overriding concern in the research and development of new materials. The copper oxides (YCBO) have excellent transition temperatures, even taking into account the fact that the working temperature of superconductors in technical applications is much lower (often less than half) than the maximum critical temperature reached. It would be very convenient if these materials had *operating* temperatures at or above the boiling point of liquid nitrogen, since liquid nitrogen is quite inexpensive and relatively easy to make, store and transport. But even for those applications requiring operating temperatures below the liquid nitrogen boiling point (77 K or $-196\,°C$), closed cycle refrigerators can bring loads (such as a sample, a device, etc.) down to 10–20 K without the complications of pouring, or in jargon 'transferring', any liquid helium into a dewar. These machines work using helium gas as a refrigerant in sealed pipes, in a way similar to a household refrigerator which employs freon or similar gases as coolants in a closed cycle system. In cases where the operating temperature of superconducting materials is in the range specified above, refrigeration means much more convenient than liquid helium already exist; thus, refrigeration in such cases might not be the main obstacle in the wide scale adoption of superconducting materials.

10.3   The outlook

Where are we heading? Gone is the facile enthusiasm which initially viewed these materials as the solution for fast transportation (maglev trains, spaceship propulsion, superconducting electric cars), fast computing (a superconducting supercomputer) and cheap generation and transport of power (superconducting transformers, power lines, and electric motors). These large scale applications are not around the corner, and perhaps they won't be for quite a while. As we discovered before (Chapter 8) in these applications the advantages (reduced wasted power, magnetic levitation, etc.) of using superconductivity-based technology might be outweighed by more mundane economic considerations. In looking for clues as to how fast recent scientific discoveries might be translated into products, we might be tempted to look at what happened to low temperature superconducting materials following the scientific discoveries in the late fifties. Engineers and materials scientists succeeded in bringing several devices to the production stage in a short period of time. These products included large magnets, SQUIDs, wires, etc.; however, their design, fabrication and maintenance costs still prevent some of them even today, thirty years after their invention, from being utilized on a much larger scale.

A somewhat more complicated situation exists for high temperature materials. Some of their characteristics are excellent, such as the high transition temperature and the very high critical magnetic fields, often exceeding 50 tesla at liquid helium temperature (commercially available superconducting magnets based on conventional materials reach magnetic fields typically one fourth or less of many of the high temperature materials when operating at liquid helium temperatures). Other properties constitute a drawback, such as the granular texture, the weak flux pinning and the brittleness of most of these materials, as discussed above in more detail. The end result, from a practical standpoint, is low current density capability and brittleness. Notwithstanding the remarkable progress achieved in these last two years in the fabrication of high T_c wires, we can justifiably expect that these drawbacks will keep engineers busy for some time.

After the first euphoria, scientists and engineers have quietly started to build and test prototypes of small devices and gadgets: bolometers, SQUIDs, liquid nitrogen 'dip-sticks', and a few other sensors and electronic components. Some of these devices are not yet up to the level of performance which can be obtained with low temperature materials, although, considering the difficulties one encounters in working with these materials, encouraging progress has indeed been made. Workers in the field already talk about very fast electronic switches and superconducting transistors driven by changes in the magnetic flux. Aside from thin films, progress has been considerably slower in making and shaping bulk super- conductors, and in developing characteristics which can be successfully used in large scale applications. While it is reasonable to expect to have several electronic devices using the new materials sometime soon, high temperature superconductivity wires will not be used in high field magnets, trains or ships for a while.

On a positive note, the discovery of high temperature super- conductors has not eclipsed research and development in low temperature superconductivity. On the contrary, there appears to be a reawakened interest in studying and exploiting low tem- perature materials, as the presence of more low temperature superconducting companies at scientific meetings and trade shows testifies to this.

It is hard to say what the next big contribution will be. Even though funds for research and development have not been forthcoming at the level that had been expected when the high T_c materials were first discovered, there are still many scientists and engineers in laboratories or at computer terminals using different ideas to approach the many problems that high temperature superconductivity has brought to us. One of them or, perhaps, many of them, will tomorrow or ten years from now probably make a significant step forward in understanding superconductivity or in making something out of superconducting materials. Judging from the history of the discoveries in superconductivity that we have just witnessed, we would be ill-advised to make predictions about what it will be and when it will occur.

CONCLUSION

Clearly, we have only indirectly answered the question which appears in the title of this book: is a new technological revolution coming? Having witnessed in recent times the invention of the transistor and the changes in our society brought about by the use of miniaturized electronic circuits (consumer electronics, computers, instruments for medical diagnosis, etc.), we wonder if comparable changes will occur with superconducting materials.

As we pointed out, it is hard to predict, at present, *which* products based on superconducting materials will be mass-produced. In fact, important technical and scientific issues regarding high T_c superconductivity have not been resolved. Economic and sociological factors sometimes stand in the way of the application of a new technology.

We have answered the question posed in the title of this book in a different way. We have looked at how important and unforeseen discoveries (vanishing electrical resistance, the Meissner effect, high temperature superconductivity) were dealt with by scientists and engineers. We have tried to give an idea of how, from a century ago up until the present day, scientific research has advanced our knowledge of superconductivity and our ability to manipulate these materials for new uses.

As to high temperature superconductivity, we hope that this

book has helped provide readers with the elements necessary to judge for themselves where these new discoveries are leading us and has aided in interpreting new developments in super-conductivity research and technology. How these discoveries will change society is something outside the realm of science and, therefore, we are much less equipped to answer this question.

BIBLIOGRAPHY

In writing this book, I used material taken from the following sources:

articles published in these periodicals: *Science, Physics Today, Historical Studies in the Physical Sciences, Review of Modern Physics, Annals of Science, Superconductor Industry, The New York Times.*

Hendrik Casimir, *Haphazard Reality*, Harper and Row (New York, 1983).

Kurt Mendelssohn, *The Quest for Absolute Zero*, Taylor and Francis (London, 1977).

Randy Simon and Andrew Smith, *Superconductors: Conquering Technology's New Frontier*, Plenum Press (New York, 1988).

Other books recently published on the discovery of high temperature superconductivity include:

Bruce Schecter, *The Path of No Resistance: the Story of the Revolution in Superconductivity*, Simon and Schuster (New York, 1989).

Robert Hazen, *The Breakthrough: The Race for the Superconductor*, Summit Books (New York, 1988).

The interested reader will find an exposition of superconducting phenomena at a higher level in the following books:

Charles Kittel, *Introduction to Solid State Physics*, John Wiley and Sons (New York, 1976).

Neil Ashcroft and David Mermin, *Solid State Physics*, Holt, Rinehart and Winston (New York, 1976).

J. Ziman, *Principles of Solid State Physics*, Cambridge University Press (Cambridge, 1972).

For an elementary exposition of quantum phenomena, see:

Tony Hey and Patrick Walters, *The Quantum Universe*, Cambridge University Press (New York, 1987).

INDEX